Contemporary's

NUMBER POWER

Analyzing Data

ELLEN C. FRECHETTE

Project Editor
Kathy Osmus

Consultants
Donna Curry
Robert P. Mitchell

CB
CONTEMPORARY BOOKS
CHICAGO

Library of Congress Cataloging-in-Publication Data

Frechette, Ellen Carley.
Number Power 8 : the real world of adult math : analyzing data / Ellen Frechette.
p. cm.
ISBN 0-8092-4213-3 (pbk.)
1. Statistics. 2. Mathematics—Charts, diagrams, etc. I. Title. II. Title: Number power eight.
QA276.12.F737 1992
519.5—dc20 92-28285
CIP

Published by Contemporary Books, Inc.
Two Prudential Plaza, Chicago, Illinois 60601-6790
(312) 540-4500
Manufactured in the United States of America
International Standard Book Number: 0-8092-4213-3

Published simultaneously in Canada by
Fitzhenry & Whiteside
195 Allstate Parkway
Markham, Ontario L3R 4T8
Canada

ACKNOWLEDGMENTS

Table on page 13 adapted from *The World Almanac & Book of Facts*, (1991 edition), © 1990 by Pharos Books, New York, NY 10166.

Excerpt on page 19 from *On an Average Day in the United States* by Tom Heymann, © 1990 by Thomas N. Heymann. Reprinted by permission of Ballantine Books, a division of Random House, Inc.

Excerpt on page 22 from *Innumeracy*, © 1988 by John Allen Paulos. Reprinted by permission of Hill and Wang, a division of Farrar, Straus & Giroux, Inc.

Graph on page 42 adapted from the U.S. Bureau of the Census, *Statistical Abstract of the United States: 1991* (111th edition), Washington, DC, 1991

Graph on page 44 adapted from the U.S. Bureau of the Census, *Statistical Abstract of the United States: 1991* (111th edition), Washington, DC, 1991

Chart on page 67 adapted from the U.S. Bureau of the Census, *Statistical Abstract of the United States: 1991* (111th edition), Washington, DC, 1991

Chart on page 76 adapted from *The Day America Told the Truth* © 1991 by James Patterson and Peter Kim. Used by permission of the publisher, Prentice Hall Press, a division of Simon & Schuster, New York.

Data on page 80 adapted from *The World Almanac & Book of Facts*, (1991 edition), © 1990 by Pharos Books, New York, NY 10166.

Graphs on page 88 reprinted by permission of *The Wall Street Journal*, © 1991 Dow Jones & Company, Inc. All rights reserved worldwide.

Graphs on page 158 © 1992 by The New York Times Company. Reprinted by permission.

Graphs on page 169 from *The World Almanac & Book of Facts*, (1992 edition), © 1991 by Pharos Books, New York, NY 10166.

CONTENTS

TO THE STUDENT

Do the statistics in the following sentences sound familiar?

- There's a 40% chance of rain.
- The unemployment rate is 7%.
- 3 out of 4 people prefer brand X.
- $3 million will be added to the district education budget.

Do you sometimes wonder what these numbers really mean?

Then welcome to *Number Power 8: Analyzing Data*. With this book, you'll have a chance to learn about the data and statistics that you see and hear about every day—on the job, in the news, and on television.

Some people become uncomfortable when faced with statistics. This may be because the language of mathematics uses symbols, expressions, and graphic materials that are sometimes unfamiliar. With this book, you will learn the language of data and statistics so that you can analyze and understand numbers.

As you work through *Number Power 8*, pay more attention to the numbers around you. As an experiment, pull out one page of any newspaper and circle all the numbers you find in the articles, advertisements, and editorials. Think about how these numbers are interpreted and how *you* might interpret them differently. Remember that although numbers may be based on facts, the statements made about them are not *always* accurate. With a basic understanding of data and statistics, you will be able to interpret data and judge other people's interpretations as well.

Good luck.

PRE-TEST

The following problems represent skills that you will learn in this book. The test measures your ability to understand and analyze different types of data. Take your time and answer the questions as completely as you can.

Problems 1–4 refer to the following data.

State	1990 Population (to the nearest million)
Alabama	4,000,000
Arkansas	2,000,000
California	30,000,000
Idaho	1,000,000
Pennsylvania	12,000,000

1. Which of the states listed has the largest population?

(1) Idaho

(2) California

(3) Pennsylvania

2. Complete the following sentence:

The table above shows the 1990 ________________ to the nearest million for five states.

3. Write a sentence that compares the population of Arkansas to the population of Alabama.

____

____

4. How many more people lived in California than in Alabama in 1990?

(1) 32,000,000

(2) 28,000,000

(3) 26,000,000

Problems 5–8 refer to the following data.

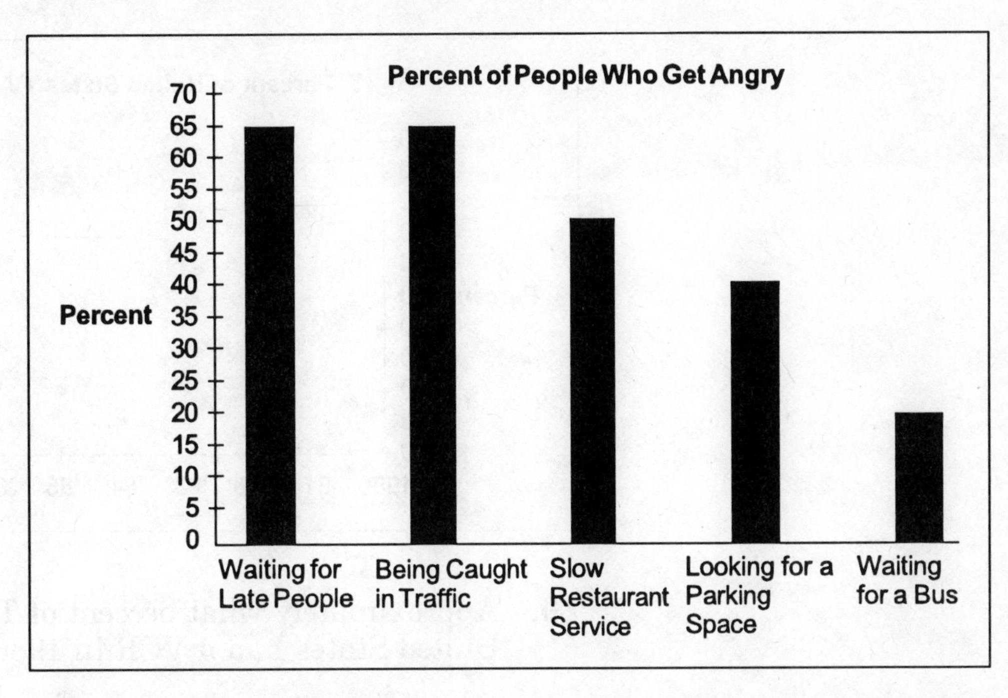

Source: *Almanac of the American People,* copyright 1988 by Tom and Nancy Biracree. Reprinted with permission of Facts on File, Inc., New York.

5. Approximately what percent of people get angry looking for a parking space?

(1) 40%

(2) 50%

(3) 55%

6. Which activity made more people angry: being caught in traffic or waiting for a bus? ____________

7. Write a statement about the percent of people who get angry at slow restaurant service. ____________

____

8. Find the mean (average) percent of people who get angry in the five situations on the graph. ________

Problems 9–12 refer to the following data.

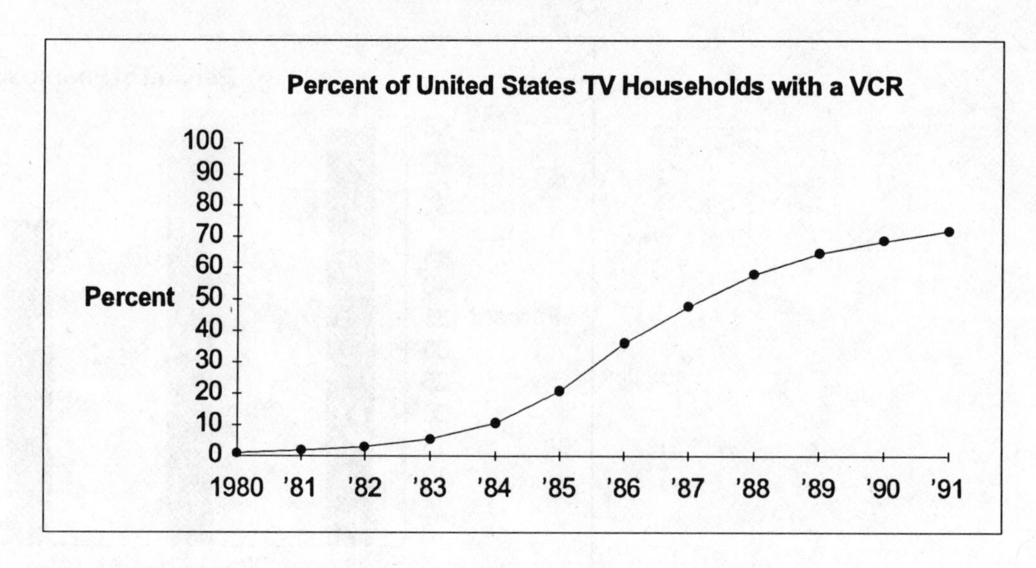

9. Approximately what percent of TV households in the United States had a VCR in 1986?

(1) 35%

(2) 30%

(3) 25%

10. Estimate and compare the percent of U.S. TV households with VCRs in 1986 and 1991.

In 1991, approximately ________ as many U.S. households had VCRs as in 1986.

(1) half

(2) $\frac{2}{3}$

(3) twice

11. If the trend continues, what percent of TV households in the United States will have a VCR by the year 2000?

(1) less than 70%

(2) about 70%

(3) more than 70%

12. Which of these statements is *not* based on the graph?

(1) More people in the United States had a VCR in 1990 than in 1989.

(2) People rented more videotapes in 1990 than in 1989.

(3) More than $\frac{2}{3}$ of all TV households in the United States had a VCR in 1991.

Problems 13–15 refer to the following data.

According to the U.S. Department of Housing and Urban Development, 40% of homeless people in 1992 were families. Single men made up 46% of the homeless population, and 14% of the homeless were single women. These statistics provide more evidence of the changing face of America's homeless. . . .

13. Complete the circle graph below. Be sure to include all labels and numbers for each segment of the graph.

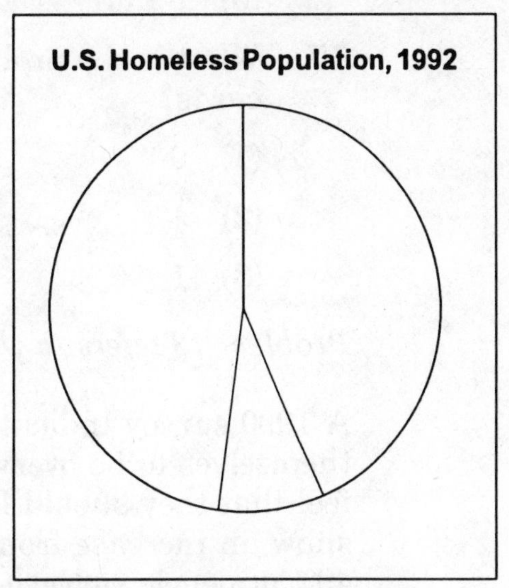

Source: U.S. Department of Housing and Urban Development

14. If a city counted 1,200 homeless people, how many of them would be single women?

(1) 168

(2) 480

(3) 8,571

15. Write a statement that compares two segments on the circle graph.

___

___

Problems 16–17 refer to the following situation.

1 2 3 6 10

Suppose the cards at left were lying facedown on a table.

16. Suppose you pick up one card. What is the probability that the card you chose is a multiple of 2?

(1) 1 in 2

(2) 2 in 3

(3) 3 in 5

17. What is the probability of choosing a 5 from these cards?

(1) 0

(2) $\frac{1}{2}$

(3) 1

Problem 18 refers to the following information.

A 1990 survey indicated that most Americans consider themselves to be overweight. A total of 78% of people surveyed feel that they should lose at least 10 pounds. These figures show an increase from previous years. In 1960, for example, 49% of people surveyed felt that they were 10 pounds overweight; in 1970, the figure was 59%; and in 1980, the figure rose to 72%.

18. Use the data above to finish constructing the line graph below. Be sure to include a title, labels on both axes, and the data line connecting the plotted points.

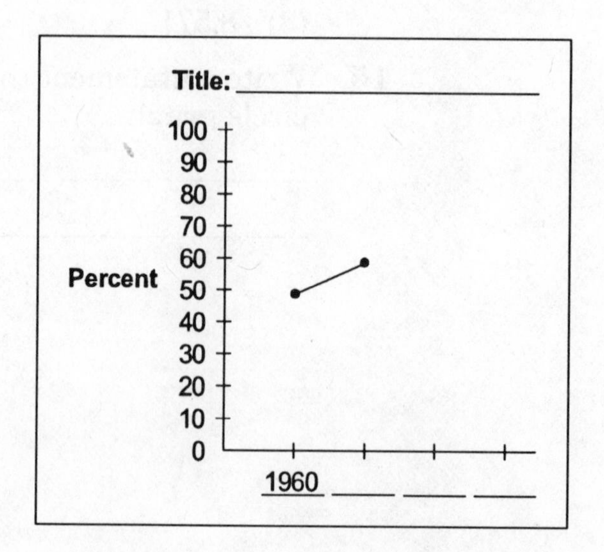

Problems 19–20 refer to the following scatter diagram.

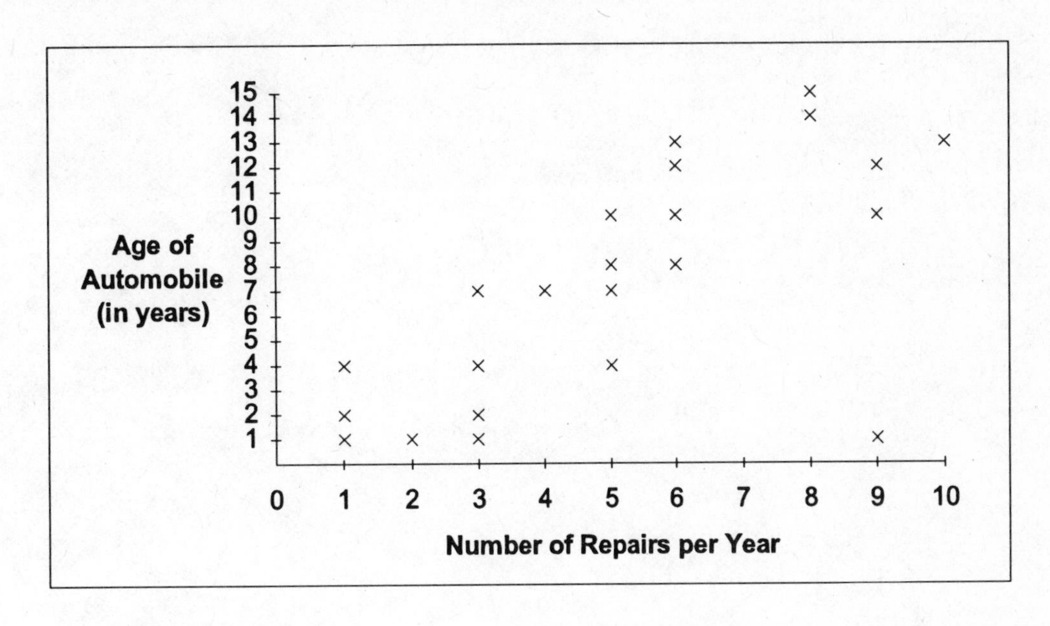

19. Is there a correlation between the age of an automobile and the number of repairs per year?

(1) yes

(2) no

20. Write a statement comparing the age of automobiles and the number of repairs per year.

____

____

▶ Answers are on pages 175–176.

PRE-TEST SKILL ANALYSIS

Problem	Skill	Chapter
1, 2, 4	Tables and Charts	2
3, 7, 15	Making a Statement About Data	1
5, 6	Bar Graphs	2
8	Mean, Median, and Mode	2
9, 18	Line Graphs	2
10	Estimation	3
11	Seeing Trends	3
12	Using Only the Information Given	3
13, 14	Circle Graphs	2
16, 17	Probability	4
19, 20	Understanding Correlation	5

CHAPTER ONE

Understanding Data

DEFINING DATA

Let's start our discussion of data with questions about you.

- What is your age?
- How many people are in your immediate family?
- How many people are living in your home?
- How many languages do you speak?
- How many times per week do you read the newspaper?
- How many different sports do you watch regularly on TV?

Data: a group of numbers that gives information about a subject

The numbers you gave in answer to the questions above represent **data**. *Data* refers to a collection of numbers that gives us information about a subject—in this case, you!

Here are some examples of data that are common in everyday life. Put a check next to any you are familiar with.

☐ The nutrition label on a cereal box gives data on calories, fat content, vitamins, and sodium.

NUTRITION INFORMATION

SERVING SIZE:	$1\frac{1}{2}$ oz. (43 g)
SERVINGS PER PACKAGE:	1
CALORIES	160
PROTEIN, g	4
CARBOHYDRATE, g	33
FAT, TOTAL, g	3
UNSATURATED, g	3
SATURATED, g	0
CHOLESTEROL, mg	0
SODIUM, mg	40
POTASSIUM, mg	125

☐ A graph in a newspaper displays population data.

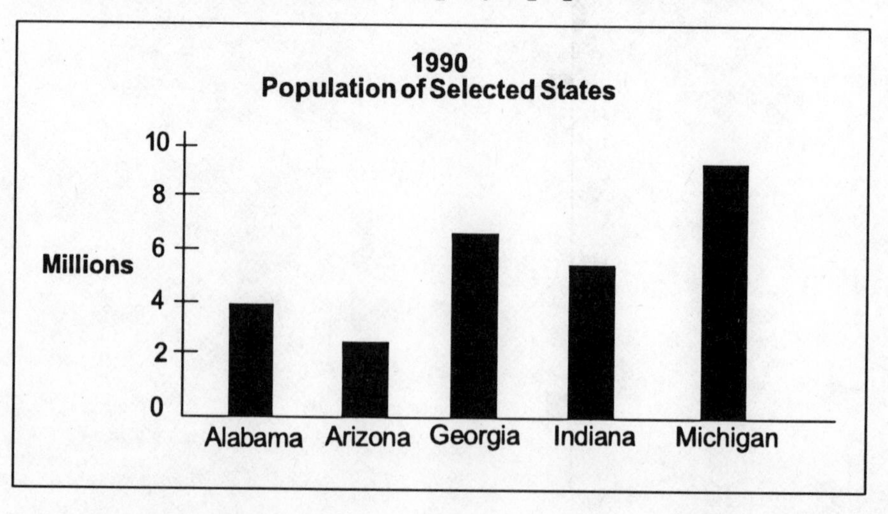

Source: U.S Bureau of the Census

☐ A document for factory employees contains data about the sizes of certain manufactured items.

Identification Number	Size
1A714	12 mm
1A820	14 mm
2B640	20 mm
2B724	26 mm
3C438	30 mm

☐ An article in a magazine provides data concerning the use of recycled products.

. . . A recent survey of Rand County residents conducted by Environment Aware, Inc., revealed that, of those surveyed,

- 47% buy and use recycled products
- 90% would buy more recycled products if the prices were lower
- 76% have trouble finding recycled products

People use data all the time to analyze situations and make decisions. Chances are that you are familiar with at least one of these examples of data.

Gathering Data

Where do we get data? How is it collected? Depending on the subject, data can be gathered in many different ways. For example:

- A scientist might collect data by measuring the quantities of chemicals in a series of test tubes.
- A teacher might collect data about test scores by keeping a record of all test and quiz grades given in a class.
- An environmental worker might gather data by observing and recording what kinds of animals and birds live in a certain region.
- A government worker might collect data by analyzing questionnaires filled out by citizens.

In fact, you may not realize it, but various types of data are being collected around you every day. If you have ever filled out a census form or responded to a telephone survey, you have supplied data that some person or organization is collecting.

Check off any of the forms of data collection that you have participated in:

☐ questionnaire
☐ consumer product survey
☐ credit or job application form
☐ U.S. Census form
☐ opinion poll
☐ voting
☐ other (What was it? ______________________________)

▶ **EXERCISE 1**

To get a better feel for data, collect your own data in this activity.

Choose a street corner near where you live and record information about the first 10 people you see, using the chart below. Or, you may choose to observe the first 10 characters you see on a TV program. To record the data when you see a person, put a slash, or **tally mark** (/), in the columns that apply. Tally marks help you "see the numbers."

Hint: When you have recorded four tally marks in a row, your *fifth* mark should be a slash through the four marks, like this:

~~////~~

This tallying method makes it easier to count up your marks later.

For example, if the first person you saw was a frowning man wearing glasses and carrying a shopping bag, your chart would look like this:

Male	Female	Wearing Glasses	Not Wearing Glasses	Smiling	Not Smiling	Carrying Something	Not Carrying Anything
/		/			/	/	

MAKING STATEMENTS ABOUT DATA

Fall 1988 Enrollment in Full-Time Day Schools

State	Pupils per Teacher	Teachers' Average Pay
Alabama	18.7	$25,500
Alaska	17.0	43,153
California	22.7	38,996
New Jersey	13.6	36,030

Source: National Center for Education Statistics

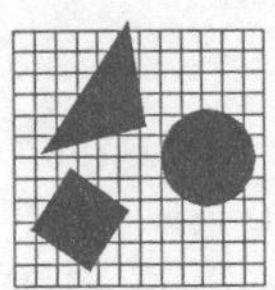

Write a short statement using the data in the chart. There is no single correct answer, so feel free to be creative. The first one is started for you.

1. California has ______________________________
______________________________.

2. ______________________________
____

What can be said about the information in the chart? Here are some statements you might have come up with:

- California has more pupils per teacher than Alabama.
- Of the four states listed, Alaska has the highest average teacher's pay.
- Alabama teachers have the lowest average pay of the four states listed, but not the fewest pupils per teacher.
- Teachers' salaries vary from state to state.

Some of these statements refer to specific pieces of information in the data.

Example: California has more pupils per teacher than Alabama.

Other statements make a more general comment about the data.

Example: Teachers' salaries vary from state to state.

The one thing that these statements have in common is that they all say something *true* about the data.

Throughout this book, you will be looking at different types of data and making statements about them. Don't worry if you had trouble coming up with a statement about the data in the table. You will get plenty of practice. By the end of this book, you will be able to *interpret and write statements* about almost any data you come across in school, at work, or in a newspaper or magazine.

IMPORTANT!

Whenever you make a statement about data, be sure to write a **complete sentence**. The phrase *seventeen pupils per teacher* does not have any meaning. However, *Alaska averaged seventeen pupils per teacher in 1988* **does** have meaning.

▶ **EXERCISE 2**

Practice writing statements about the data you collected on page 12. The questions below should help you get started. Remember that you can be as straightforward or creative as you like when writing your statements. Just be sure that each statement is a complete sentence and is true according to the data you collected.

Questions to help you get started:

- Out of 10 people, how many were female?
- Did you see more people with glasses or without?
- How many people were carrying something?
- Who was smiling more often—men or women?
- Did more men than women wear glasses?

Statement 1: ____

____

Statement 2: ____

____

Statement 3: ____

____

Statement 4: ____

____

Statement 5: ____

____

WHAT ARE STATISTICS?

Of the 4,280 people who responded to our magazine's survey, 71% said that the government should provide subsidized child care for all workers. Twenty percent stated that child care is not the government's concern, and 9% had no opinion on the issue. Furthermore . . .

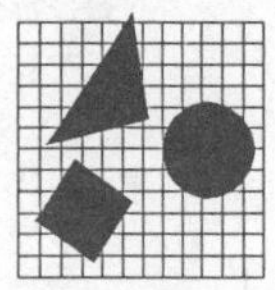

- What numbers appear in this article?
- Have you read articles similar to this?
- How do these numbers differ from the data you saw earlier?

Statistics: summaries of data

Many statements you read in newspapers and magazines or hear on the radio or television contain **statistics** such as those in the article above. The term *statistics* refers to the collection, organization, and interpretation of *data.*

For example, in Exercise 1, you collected data about a group of people. If you organized and interpreted this information, you might use statistics to help present the facts.

Imagine what the article above would sound like if it simply listed the information obtained from the survey:

QUESTION: Do you think that the government should provide subsidized child care for all workers?

Miriam Brown from Spokane, WA, responded yes.
Carlos Young from Cairo, IL, responded yes.
David Antony from Los Angeles, CA, responded no.
Walter Fireton from Baton Rouge, LA, responded yes.
Vera Roundtree from Orono, ME, responded no.
Mary Caviness from Kansas City, MO, had no opinion.
Maureen Burns from Windham, NH, responded yes.

As you can see, an article that listed each person's opinion would be extremely long. And it would be practically impossible to understand the results of the survey. Instead, the writer **summarizes** the data by using percents:

71% said "yes."

20% said "no."

9% said "no opinion."

■■ YOU TRY IT ▶▶

Using tally marks, record the responses given on page 15 to help you "see the numbers." Miriam Brown's response has already been recorded.	YES	NO	NO OPINION
	/		

▶ Answers are on page 176.

How Are Statistics Presented?

Once a group of data is collected, the numbers are counted and put into a form that is easy for people to understand.

In the example on page 15, the magazine probably issued a questionnaire or did a phone survey to determine what people thought about government-subsidized child care.

- What different ways can you think of to organize the information?
- How did the magazine present its data?

●● MATH RECAP ▶▶

Data can be represented by **percents**. Remember that *percent* means "out of one hundred."

The statement that 71% think government should provide child care means that 71 out of every 100 people surveyed think government should provide child care.

The magazine received 4,280 responses to its survey. It chose to represent its data as *percents*. What other methods could it have used?

1. It could have listed each individual's response separately, as you saw on page 15. This is a long and awkward method.
2. It could have given *numbers* instead of percents:

   "Out of 4,280 responses, 856 people did not like the idea of government-subsidized child care."

3. It also could have used *fractions*:

"Approximately $\frac{7}{10}$ of respondents favored government-subsidized daycare."

4. *Ratios* are another way to present data. You may be familiar with this technique in advertisements.

"7 out of 10 (7:10) respondents favored government-subsidized daycare."

Ratios are most often written in fraction form: $\frac{7}{10}$.

▶ **EXERCISE 3** Read the following lines taken from newspapers, magazines, and work reports. Express the numbers given as a percent, a fraction ratio, and a ratio.

Example: "3 out of 10 students do not own a calculator."

30% ___ $\frac{3}{10}$ ___ 3:10 or 30:100 ___
percent ___ fraction ratio ___ ratio

1. "Employees pass the physical exam 95 out of 100 times."

_____ percent _____ fraction ratio _____ ratio

2. "8 out of every 100 people in Newton are unemployed."

_____ percent _____ fraction ratio _____ ratio

3. "7 out of 10 customers prefer orange to grape juice."

_____ percent _____ fraction ratio _____ ratio

4. "Out of 200 questionnaires returned, 60 were completed by females."

_____ percent _____ fraction ratio _____ ratio

5. "The risk of mechanical error is 1 in 100."

_____ percent _____ fraction ratio _____ ratio

6. "The ratio of computer users to non-users was 4 to 5."

_____ percent _____ fraction ratio _____ ratio

▶ Answers are on page 176.

COMPARING NUMBERS

- Your town government votes to give $80,000 to the parks and recreation department.
- You get an offer of $80,000 for your house.
- The legislature voted in favor of an $80,000 government-subsidized school lunch program.

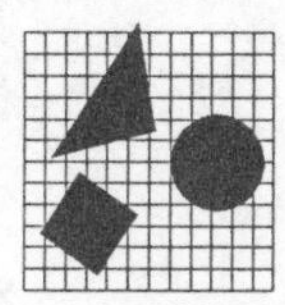

- Is 80,000 a large number or a small number?
- How do you decide?

One of the most important things to know about a number is this:

> A number is neither large nor small until you compare it to another number.

Let's look at the first example above. What other information would help you decide whether or not $80,000 is a lot of money for parks and recreation?

Here are some numbers you might compare it to:

- **The amount of money allocated the previous year.** For example, if the town had allocated $200,000 last year, perhaps $80,000 does not seem like a very large figure.
- **The total amount of money in the town budget.** If the town had a budget of $125,000, then $80,000 would seem like a lot of money, given the fact that the town must also support schools, fire and police departments, maintenance work, etc.
- **The amount allotted to parks and recreation in similar neighboring towns.** If you compared $80,000 to the $10,000, $8,000, and $22,000 that three similar towns were spending, you would certainly think $80,000 was a lot of money.

Compare: to look at two or more numbers and see their similarities and differences; also, to decide which is larger or smaller

When you see or hear a number, be careful not to make a judgment about its size until you have a chance to **compare** it to other related figures.

STATS

Did you know that on an average day in 1989 . . .
- the federal government spent $46,027,400 on education?
- the federal government spent $807,260,270 on defense?

Which of the following numbers would be useful in determining whether the $80,000 offered for your house is a large or small number? Put a check mark next to the numbers that you would use to compare with the $80,000.

☐ a. the amount paid for a similar house in your neighborhood

☐ b. your neighbor's salary

☐ c. the number of houses on your street

☐ d. the price you paid for the house last year

You were right if you chose *a* and *d*. No matter how high or low your neighbor's salary is or how many houses there are in your neighborhood, these numbers will not tell you anything about the price of your house.

Now finish the following sentences:

- $80,000 for a house would be a *large* amount of money if ______________________________.
- $80,000 for a house would be a *small* amount of money if ______________________________.

Here are some sample sentences:

- $80,000 for a house would be a *large* amount of money if **I paid only $40,000 for it last year**.
- $80,000 for a house would be a *small* amount of money if **my neighbor sold the same kind of house for $100,000 two months ago**.

■■ YOU TRY IT ▶▶

Make two statements about the $80,000 school lunch program.

1. $80,000 for the school lunch program would be a lot of money if ______________________________ ______________________________ ______________________________.

2. $80,000 for the school lunch program would be a small amount of money if ______________________________ ______________________________ ______________________________.

▶ Sample statements are on page 176.

Don't rely on other people's interpretations of the size of numbers. It's important to draw your *own* conclusions about the information. A good start is to *compare* the information being presented to other amounts. To see how the same number can be looked on as both large and small, read the following excerpts from two different newscasts:

> . . . Yesterday, a crowd of approximately 100 people marched in front of City Hall, protesting the pay raises voted in on Tuesday. A handful of people held up signs supporting the increased salaries. . . .

> . . . Compared to last year's enormous protests over high salaries for government officials, the group of 100 or so protesters in front of City Hall yesterday was hardly noticed. . . .

The *fact* is that 100 people stood outside City Hall. How you *interpret* that depends on what you compare the numbers to.

▶ EXERCISE 4 Look at the figures given below. What other numbers would be helpful in deciding whether these numbers are small or large?

Example: $100 to repair a stereo
It would be helpful to know *how much another place would charge for repairs*.

1. 400 people at a fund-raiser
It would be helpful to know ______________________________
______________________________.

2. 90 people laid off in a company
It would be helpful to know ______________________________
______________________________.

3. 3 guns confiscated in a police raid
It would be helpful to know ______________________________
______________________________.

4. $20 to purchase a painting
It would be helpful to know ______________________________
______________________________.

5. $2,000 to buy a car
It would be helpful to know ______________________________
______________________________.

▶ Sample answers are on page 176.

HOW BIG IS A BILLION?

- Ticket sales from the movie *Batman* totaled $251.2 **million** in 1989.
- The United States' federal debt (the amount by which government spending exceeded government income) in 1989 was $2.8 **trillion**.

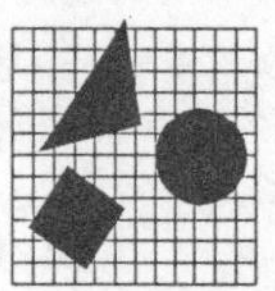

- Do you have a good understanding of the size of the numbers at left?
- When you read a very large number, what do you do to determine its size?

For many people, the bigger a number is, the less meaning it has. Let's review **place value** to help in our understanding of large numbers.

1	one
1 × 10 = 10	ten
10 × 10 = 100	one hundred
100 × 10 = 1,000	one thousand
1,000 × 10 = 10,000	ten thousand
10,000 × 10 = 100,000	one hundred thousand
100,000 × 10 = 1,000,000	one million
1,000,000 × 10 = 10,000,000	ten million
10,000,000 × 10 = 100,000,000	one hundred million
100,000,000 × 10 = 1,000,000,000	one billion

▶ **EXERCISE 5** Express the values in word or numeral forms as needed.

Examples: 30,000,000 *thirty million*

two trillion *2,000,000,000,000*

1. 900,000 ______

2. 8,000,000 ______

3. 90,000,000 ______

4. one hundred million ______

5. two billion ______

6. seven hundred thousand ______

▶ Answers are on page 176.

● ● **MATH RECAP** ▶▶ To multiply a whole number by 10, add a zero at the right.

■■ YOU TRY IT ▶▶

The function of zeros can help you understand the relative size of numbers. For example,

One million is how many times as large as a hundred thousand?

■ Write each amount in numeral form. Put the smaller number on top.	100,000 1,000,000
■ Cross out one zero at a time in each number until there are no zeros left in the smaller number. Be sure to cross out the same number of zeros in each number.	1~~00~~,~~000~~ 1,0~~00~~,~~000~~
■ Count the number of zeros left in the larger number.	1,0~~00~~,~~000~~ How many zeros are left? ____
■ Your knowledge of place value tells you that one zero represents 10.	One million is *ten* times as large as one hundred thousand.

How many times larger is one billion than one million?

3 zeros left → 1,~~000~~,~~000~~ ← cross out 6 zeros
1,000,~~000~~,~~000~~

One **billion** equals one **thousand** times one **million**.

When you come across large numbers like these, you need to have a basic understanding of their *relative* size (size in relation to other numbers). There is a big difference in value between a million and a billion! To illustrate how big a difference, think about this:

- It takes $11\frac{1}{2}$ days for a *million* seconds to pass.
- It takes almost *32 years* for a *billion* seconds to pass.

STATS

If you had five million dollars, you could spend $10,000 a day for a year and still have more than a million dollars left.

▶ EXERCISE 6 Use the "times 10" chart on page 21 and the method shown in "You Try It" on page 22 to fill in the following.

All figures below refer to the year 1991.

Example: In 1991, almost four million Roman Catholics lived in East Asia. In 1991, almost four hundred million Roman Catholics lived in Latin America.

In 1991, there were _100_ times as many Roman Catholics in Latin America as in East Asia.

4,~~000~~,~~000~~ four million

400,~~000~~,~~000~~ four hundred million

2 zeros mean 100 times

1. Approximately 160,000,000 Buddhists live in South Asia. About 160,000 Buddhists live in North America.

   There are ______ times as many Buddhists in South Asia as in North America.

2. An estimated 750,000,000 people in East Asia consider themselves atheist or nonreligious. An estimated 75,000,000 Europeans include themselves in this category.

   ______ times as many East Asians as Europeans consider themselves atheist or nonreligious.

3. One hundred thousand Jews live in Oceania. One million Jews live in Latin America.

   There are ______ times as many Jews living in Latin America as in Oceania.

4. Approximately one billion five hundred million Christians exist worldwide, compared to fifteen million Jews.

   There are approximately ______ times as many Christians as Jews worldwide.

5. Ninety million Catholics live in Africa, out of a total of 900 million Catholics worldwide.

   There are ______ times as many Catholics in the world as in Africa.

▶ Answers are on page 176.

NUMBERS VS. RATES

1992 Accident Report	
Highway	Number of Accidents
A	110
B	75
D	10
G	90
HH	12
MM	170

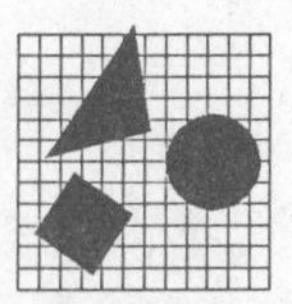

Which of the highways in the chart at left would you prefer to drive on? Why?

Did you say that you would rather drive on *Highway D* because it had the fewest accidents and therefore was the safest to drive on?

You may have made the common mistake of confusing *numbers* with *rates*. Before you can decide which highway is the safest, you need to know some other numbers.

For example, would you change your mind if you knew that only 3,000 cars travel on Highway D per year, compared to 75,000 on Highway B?

Rate: a quantity measured per unit of another quantity

The chart above really does not give you much useful information. What would be more helpful is to know *how many accidents occurred per one hundred cars*. In other words, we need to know the **rate** of accidents, not just the **number** of accidents.

Now look at the revised chart. Unlike the chart above, this chart gives enough information to determine the *rate of accidents*. Do you know how to determine the rates?

Highway	Number of Accidents	Total Number of Cars per Year
A	110	55,000
B	75	75,000
D	10	3,000
G	90	60,000
HH	12	8,000
MM	170	80,000

You Try It ▸▸

To figure out each rate, divide the number of accidents by the total number of cars.	Highway A: 110 ÷ 55,000 = ?
A calculator entry would look like this:	1 1 0 ÷ 5 5 0 0 0
The calculator displays:	.002
Change the decimal to a percent by moving the decimal point two places to the right.	.002 = .2%
Now find the accident rates for the remaining highways. Which highway has the lowest accident rate? ____	**Highway**

Highway	Decimal		Percent
A	.002	=	.2 %
B	____	=	____ %
D	____	=	____ %
G	____	=	____ %
HH	____	=	____ %
MM	____	=	____ %

▸ Answers are on page 176.

Highway B, with a rate of .1%, is the safest. Although this highway had 75 accidents, its accident *rate* is relatively low considering that so many cars travel on it.

Remember: A *rate* gives you more information than a number does, because a rate is actually *two* numbers—a number *compared to* another number.

A percent is the rate most commonly used—it represents a number out of one hundred. For example, which statement below gives you more information?

Statement A There are 6,400 people living below the poverty level in our town.

Statement B In this town, 17% of the people live below the poverty level.

Both statements are true, but statement B is the only statement that gives you information about the *size* of the poverty problem. Unless you know the total number of residents, statement A does not tell you much. The number 6,400 could be a huge number (if residents totaled 12,000) or a relatively small number (if residents totaled 220,000).

STATS

There were **1,820** murders in New York City in 1989. There were **432** murders in Houston, Texas, the same year.

BUT

The murder **rate** in New York City was **26** per 100,000 people. In Houston, the murder **rate** was **27** per 100,000 people.

▶ EXERCISE 7 Part One

Look at the information below, then make

- two statements about *numbers*
- two statements about *rates*
- a statement comparing numbers vs. rates

. . . A survey organization asked the employees of five different companies in the metro area this question: "Are you satisfied with your company's health benefits?" The results are shown below:

Company	Satisfied Employees	Total Number of Employees
Walton Company	190	1,000
Kramer, Inc.	13	87
Psytech	440	3,800
Ranger, Intl.	90	450
Horton & Co.	26	105

Listed below are some questions to help you get started.

1. What company had the *smallest number* of satisfied employees, and what is that number?

   Statement: ______________________________

   ____

2. What companies had the *largest numbers* of satisfied employees?

   Statement: ______________________________

   ____

3. What company had the *lowest rate* of satisfaction?

   Statement: ______________________________

   ____

4. What company had the *highest rate* of satisfaction?

   Statement: ______________________________

   ____

5. Compare the *highest number* of satisfied employees with that company's employee satisfaction *rate*.

   Statement: ______________________________

   ____

Part Two

One of the most common rates that you will find in the news is the *unemployment rate*. Instead of reporting exactly how many people are out of work, the government reports a rate.

Labor Force: that part of the population that is willing and able to work

A **6.9%** unemployment rate, for example, means that **6.9** (almost **7**) people out of every **100** people in the **labor force** are not working.

The government also provides statistics for different segments of the labor force. For example, in 1991, the unemployment rate for the entire labor force was 6.6%. However, among young people aged 16–19 the rate was 18.6%—nearly three times that rate!

Use the chart below to write three statements about the unemployment rates for various years and among various segments of the labor force.

U.S. Unemployment Rates

Year	Total Work Force	Both Sexes 16–19 Years	White	Black
1984	7.4%	18.9%	6.5%	15.9%
1986	6.9	18.3	6.0	14.5
1988	5.4	15.3	4.7	11.7
1990	5.4	15.5	4.7	11.3
1992*	7.0	18.3	6.2	13.7

*January figures

Source: Department of Labor, Bureau of Labor Statistics

1. Compare the unemployment rate for the total work force in two different years.

___

___

2. Write a statement about unemployment among teenagers in the United States using the figures above.

___

___

3. Compare the unemployment rates of white Americans to the rates of black Americans during one year.

___

___

▶ Sample statements are on page 176.

A Look at the U.S. Census

Every 10 years, the federal government collects, tabulates, and publishes statistics about the American people. These statistics make up the **U.S. Census**. The census is used, among other things, to determine

- how many representatives each state should have in Congress
- how much money each city and state should receive in federal aid
- how much money to allocate for bilingual education
- the need for low-income housing

The census attempts to count *each and every person* living in the United States. Just imagine what a data collection task that is—considering that there are almost 250 million people living in the United States today! To get information from all these people, the government sends out questionnaires to home addresses. Those people who do not respond to the questionnaires are reached by door-to-door census takers. An estimated 315,000 civil servants were needed to collect and process the data for the 1990 census.

Because the census is such an enormous project, some people slip through the cracks and are not counted. This results in an **undercount**—a number below the actual total.

What did the 1990 census tell us about the people of America? Here are some statistics:

Age

The nation's median age is 32.9 years. This is up from 30 years in 1980.

Sex

51% of the population is female. This means that there are 6.2 million more women than men in America.

Households
55% of U.S. households are maintained by married couples. This is down from the 1980 figure of 60%.

Persons living alone accounted for 25% of households.

Where We Live
77.5% of the U.S. population lives in metropolitan areas.

The Los Angeles/Anaheim/Riverside metropolitan area experienced a 26% population growth, while the New York/New Jersey/Long Island/Connecticut area experienced a 3% growth.

▶ EXERCISE 8

The chart below contains some more data from the U.S. Census. Use the data to answer the following questions.

City	Official Population Count	Undercount
New York	7,322,564	3.0%
Los Angeles	3,485,398	5.1
Chicago	2,783,726	2.6
Houston	1,630,553	5.0

Source: Copyright 1991 by *USA TODAY.* Reprinted with permission.

1. Which city has the highest official population count?

   ____

2. Which city has the highest percent of people who were not counted (undercounted) in the census?

   ____

▶ Answers are on page 177.

CRITICAL THINKING

Census data can be collected by mail, phone, and door-to-door interviews. Which person would be less likely to be counted: a homeless person or a homeowner? What effect do you think this has on the census statistics?

GROUP PROJECT

Collecting Data

Use the questionnaire below to collect data. The more people in your group collecting the data, the better.

- You may photocopy the questionnaire and have people fill it out themselves, or you may survey people directly by asking them the questions out loud and recording their responses.
- Try to get information from a variety of people—students, co-workers, friends, and some people whom you don't know well. Try to get a variety of ages and ancestry if possible.
- Keep the responses to your survey in a folder, because you will be using the data you collected as you work through this book.

Questionnaire/Survey

Sex M ☐ F ☐ Age ☐

Are you married? Yes ☐ No ☐

Do you consider yourself more of a "day person" or a "night owl"? ____________

What section of the newspaper do you read first? ____________

Do you live alone? Yes ☐ No ☐

How many hours of television do you watch per week? ____

Do you have a dog? ☐ a cat? ☐ a bird? ☐

Do you recycle any products? Yes ☐ No ☐

If yes, paper? ☐ aluminum? ☐ glass? ☐

Are you satisfied with the current president? Yes ☐ No ☐

Which are you more concerned about, the economy or the environment? ____________

Should Americans buy only products made in America? Yes ☐ No ☐

How many close friends do you have, not including family? ____

How many times per week do you see people for a social occasion? ____

Count up the responses to your survey using the space below. First, use tally marks as you did on page 12. Then count up the tally marks and write a total in the blank provided.

Sex
Male: ____________
Female: ____________

Age
15–24: ____________
25–34: ____________
35–44: ____________
45–54: ____________
55–64: ____________
65–74: ____________
75–84: ____________

Married
Yes: ____________
No: ____________

"Day Person" or "Night Owl"
Day: ____________
Night: ____________

Section of Newspaper (write in)
____________ ______
____________ ______
____________ ______

Living Alone
Yes: ____________
No: ____________

Hours of Television per Week
0–4: ____________
5–8: ____________
9–12: ____________
13–16: ____________
17–20: ____________
Over 20: ____________

Pets
Dog: ____________
Cat: ____________
Bird: ____________

Recycling
Paper: ____________
Aluminum: ____________
Glass: ____________

Satisfied with President
Yes: ____________
No: ____________

Biggest Concern
Economy: ____________
Environment: ____________

Buy American
Yes: ____________
No: ____________

Number of Close Friends
0–2 ____________
3–5 ____________
6–7 ____________
8–9 ____________
10 or more ____________

Social Occasions per Week
0–2 ____________
3–5 ____________
6–7 ____________

CHAPTER TWO

Organizing Data

TABLES AND CHARTS

How Salty Snacks Stack Up
(One-Ounce Serving)

	Potato Chips	Pretzels	Popcorn*	Corn Chips	Peanuts
Calories	152	111	76	153	170
Fat	10 g	1 g	1 g	9 g	14 g
From Fat	60% of calories	8% of calories	12% of calories	53% of calories	74% of calories
Sodium	160 mg	451 mg	1 mg	218 mg	138 mg

*Air-popped corn. The microwave type has about twice the calories. It contains nine grams of fat (that's 55 percent of its calories) and 196 milligrams of sodium.

Source: Copyright 1991 by Consumers Union of United States, Inc., Yonkers, NY 10703. Reprinted by permission from *Consumer Reports*, June 1991.

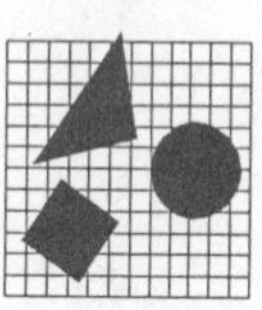

1. Why are tables like the one above often used to display data?
2. What does this table have in common with other charts you've seen?

Effective tables and charts have at least four elements:

- A *title* describes the subject of the table or chart.
- *Headings* (**column** and **row** labels) show what information will be provided in the table.
- *Data* (for example, the numbers listed under and next to the headings above) gives the most specific information in the table.
- The *source* is usually found in smaller print below a graph, table, or chart, and it tells where the data actually came from.

Column: a group of numbers or words listed vertically

Row: a group of numbers listed horizontally

To find specific information in a table, you will need to read across a row and down a column to where the row and column intersect (meet).

Look at the table on page 34. How much sodium is in one ounce of corn chips?

- Find *Sodium* in the list of row headings at the left in the table.
- Follow the row across until you come to the *Corn Chips* column, labeled at the top of the table.
- The number at the intersection of the *sodium* row and the *Corn Chips* column is 218 mg. There are **218 mg** of sodium in one ounce of corn chips.

Charts and tables are useful because they organize data in an easy-to-read format.

▶ EXERCISE 1

Part One

Use the table on page 34 to fill in the blanks.

1. One ounce of air-popped popcorn contains ______ gram(s) of fat.

2. ______% of the calories in peanuts come from fat.

3. One ounce of microwave popcorn has about ______ calories.

Use the table to decide if each statement is true or false.

T F **4.** Corn chips have more calories than potato chips.

T F **5.** Popcorn has 4% more calories than pretzels.

T F **6.** A bowl containing 1 ounce of corn chips and 1 ounce of pretzels contains more than 200 calories.

Make two statements using the data in the table. The first one has been started for you.

7. A one-ounce serving of peanuts ______________________

______________________________________.

8. ______________________________________

____

▶ Answers are on page 177.

CRITICAL THINKING

Consumer Reports is a magazine published by a nonprofit, independent organization. Do you think the information in its charts would be different from charts you might see in advertisements? Why?

Part Two

Use the information given in this newspaper article to fill in the table below.

Step 1. Read the article below.

Step 2. Write a title for the table and fill in the source line.

Step 3. Label the columns (*$50,000 or More* has been filled in) and rows (the *Bank Credit Card* row has been labeled).

Step 4. Fill in the data (*94%* has been filled in).

. . . According to *Money* magazine, 94% of Americans earning $50,000 or more had bank credit cards, and 54% had gasoline credit cards. Fourteen percent of this group had an American Express Gold card. Of Americans earning $25,000–$34,999, 73% had a bank credit card, 33% had gasoline cards, and 3% had an American Express Gold card. Finally, of Americans earning less than $15,000, only 36% had a bank card, and only 1% had an American Express Gold card. However, 19% of this group had gasoline cards. . . .

Title ______________________________

	Income Level		
Type of Card	$50,000 or more		
Bank Credit Card	94%		

Source: ______________________________

▶ Answers are on page 177.

Part Three

Use the table on page 36 to fill in the blanks.

1. ________% of Americans earning $25,000–$34,999 have a bank credit card.
2. 19% of Americans earning less than $15,000 have a ________ credit card.
3. 3% of Americans earning ________ have an American Express Gold card.
4. ________% more Americans earning $50,000 or more have a bank credit card than a gasoline credit card.

Use the table to decide if each statement is true or false.

T F 5. 14% of Americans earning $25,000–$34,999 hold an American Express Gold card.

T F 6. Americans earning $50,000 or more are the highest percent of credit card holders for every card listed.

T F 7. Less than 50% of Americans earning $25,000–$34,999 hold a bank credit card.

T F 8. 1% of Americans earning less than $15,000 hold an American Express Gold card.

Make two statements using the data in the table. The first one has been started for you.

9. The bank credit card ________________________________

____.

10. ________________________________

____

▶ Answers are on page 177.

BAR GRAPHS

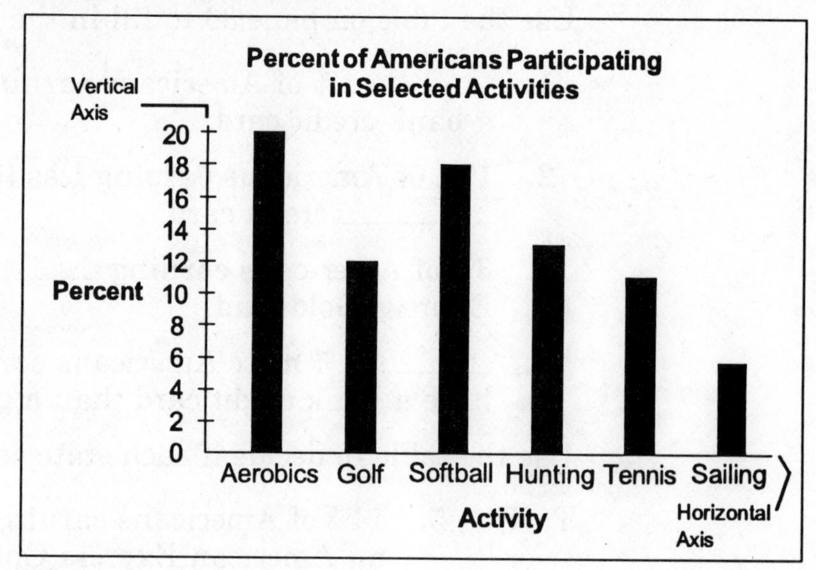

Source: Gallup Organization

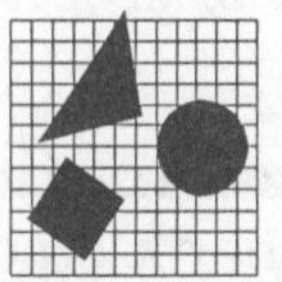

- Have you ever seen a graph similar to this one?
- What is being compared on the graph?

How do you "read" a **bar graph** like the one above? Just as with a table or chart, you need to pay attention to the title and source line. Listed below are other major parts of a bar graph that will help you interpret the data.

- *The vertical axis and the horizontal axis and their labels*

On the graph above, the **vertical axis** represents percent, while the **horizontal axis** shows different activities.

- *The bars.* Bars can be vertical (as in the graph above) or horizontal. Value is shown by the height or length of each bar.

Each bar on the graph above represents a different activity; for example, aerobics or softball.

- *The scale and range.* A **scale** marks the length of each bar and is labeled in equal increasing values. The **range** is the lowest value on the graph through the highest.

The scale on the graph above shows percents in increases of 2 with a range of 0 to 20.

You may already know how to read a bar graph. How can you find out what percent of Americans golf?

- Find the bar labeled *Golf* on the horizontal axis.
- Move your finger from the top of the bar across to the corresponding value on the vertical axis.
- The value is 12; therefore, the graph tells you that **12%** of Americans play golf.

Now figure out what percent of Americans play tennis. Can you see that the *Tennis* bar rises to a point between two points labeled on the scale?

The *range* of percents given on the graph is 0% to 20%, and the values are labeled in *multiples of two* (2, 4, 6, 8, etc.)

The value between 10 and 12 is 11; therefore, **11% of Americans play tennis**.

YOU TRY IT

Take a few minutes to determine from the graph on page 38 what percent of Americans participate in each of the activities listed.

1. Aerobics: ________________

2. Softball: ________________

3. Hunting: ________________

4. Sailing: ________________

▶ Answers are on page 177.

Bar graphs are useful for more than just showing specific data. A quick look at a bar graph allows you to make comparisons very easily.

For example, do more Americans play softball or tennis?

____

All you need to do is compare the lengths of the two bars. The softball bar is taller than the tennis bar. Therefore, **more Americans play softball**.

CRITICAL THINKING

How do you think the Gallup Organization got the data for this information? Did its staff ask every single American if he or she plays golf or tennis?

Now let's apply what you have just learned to another bar graph. Notice the similarities and differences between it and the one on page 38.

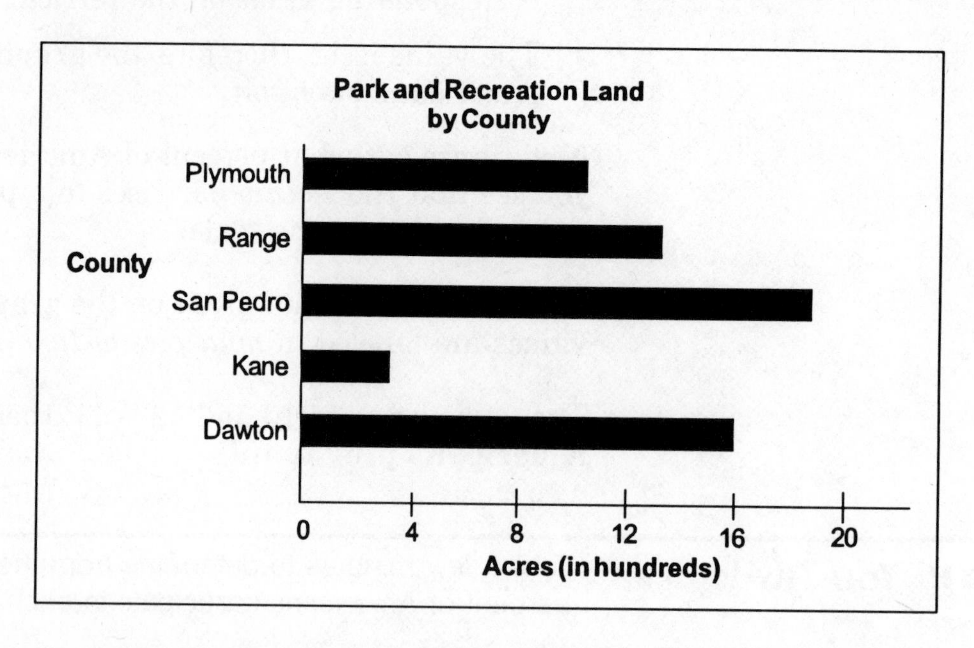

■■ YOU TRY IT ▶▶

1. What is the *title* of the graph? Title: ____________________

2. What does the *vertical axis* represent? Vertical Axis: ____________________

3. What does the *horizontal axis* represent? Horizontal Axis: ____________________

4. What is the *range* of values on the scale? Range: ____ to ____

5. The scale is measured in *multiples* of ____________________. Multiples of ____

Now use the graph to make some comparison statements. Do not look at specific data points. Instead, use just the relative lengths of the bars to make these statements.

6. Dawton County has more park and recreation land than ____.

7. Kane County ____ ____.

8. ____ ____.

▶ Answers are on page 177.

▶ EXERCISE 2 Use the information given below to construct a bar graph. Be sure to include a title, labels on both axes, and a source line.

Hint: Make the **range** of values 0 to 120. Use **multiples of 10**.

> . . . The International Atomic Energy Agency stated that in 1990 the United States had 111 nuclear reactors in operation. The agency compared this number to the 55 reactors operating in France, the 18 operating in Canada, and the 39 operating in both Japan and the United Kingdom. . . .

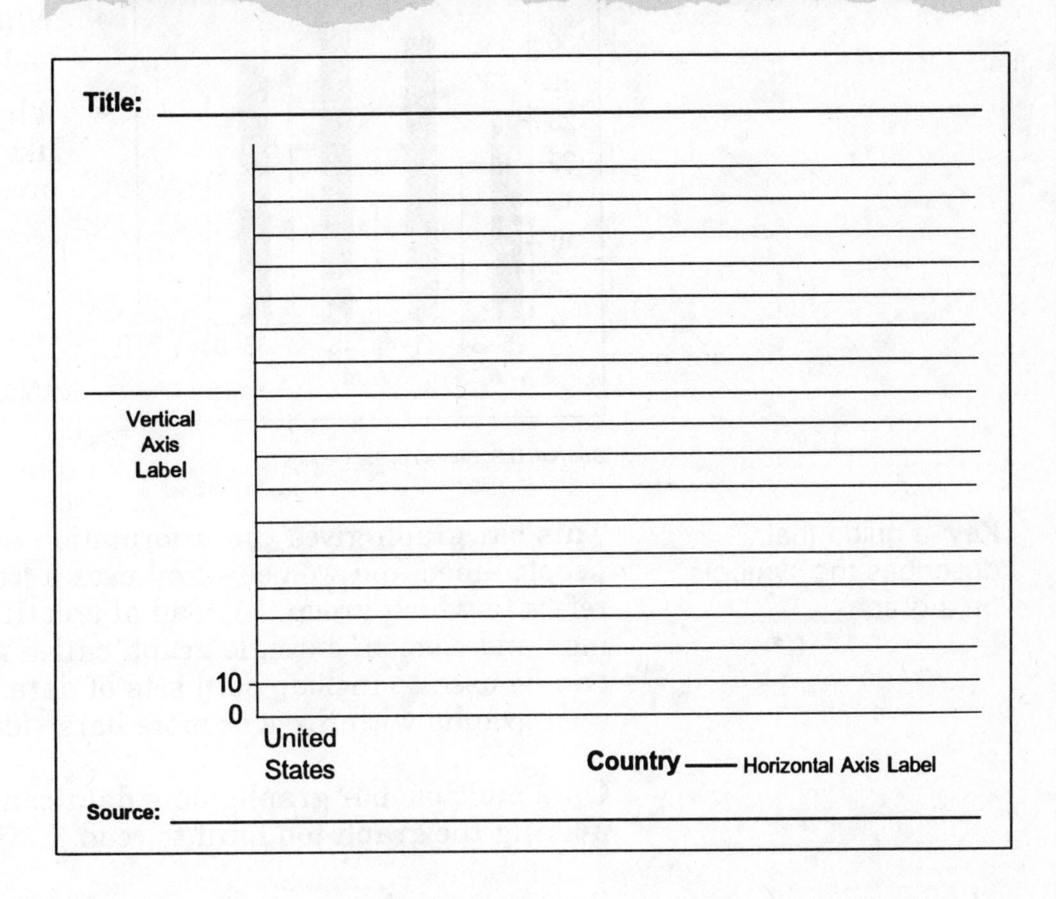

Now use the graph to make some comparison statements.

1. In 1990, the United States had ______________________________.

2. Canada had ______________________________.

3. Japan and the United Kingdom ______________________________.

▶ Sample answers are on page 177.

DOUBLE BAR GRAPHS

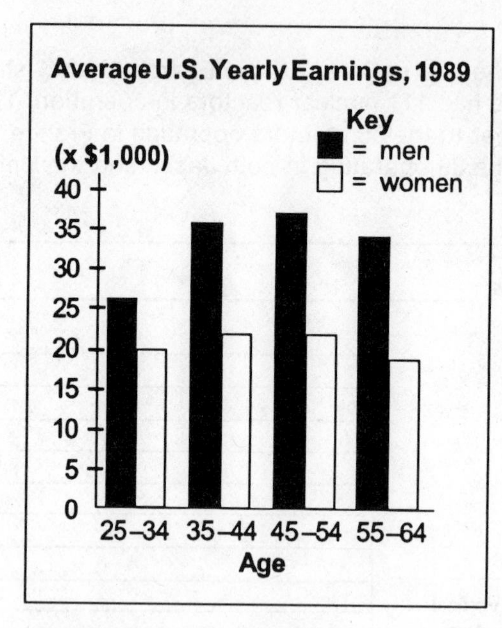

Source: U.S. Bureau of the Census

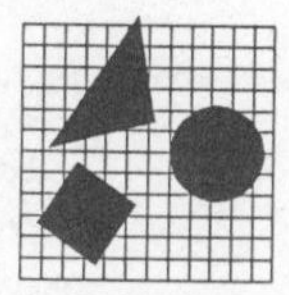

- What does the key tell you on this graph?
- Why would double bars like these be used to present data?

Key: a guide that describes the symbols on a graph

This bar graph gives you information about two groups of people—men and women—and uses a **key** to show which bar refers to which group. Instead of creating separate graphs for men and women, a single graph called a **double bar graph** can be used to include both sets of data. You may also have seen graphs with three or more bars side by side.

On a multiple bar graph, more data can be presented without making the graph too hard to read.

For example, if you were interested in finding information about *women's* average earnings, what *type* of bar would you look at? ____________________

And if you wanted to know what the average earnings were for men aged 55-64, where would you look? ____________________

Did you answer the **unshaded bars** to the first question and the **last shaded bar** to the second question?

As with all bar graphs, comparisons between groups can easily be made simply by looking at the length of the bars. In double bar graphs, comparisons can be made *within* groups and *between* groups.

For example, using the graph on page 42, you can compare average earnings of men vs. ________________ in the *same* age group.

You can also compare earnings among different ________________.

This graph makes it easy to compare average earnings of **men vs. women in the same age group** and among **different age groups.**

■■ YOU TRY IT ▶▶

1. Write a statement that compares the average earnings of one age group to those of another.

Statement 1: ________________

____

2. Now write a statement that compares the average earnings of men to those of women.

Statement 2: ________________

____

Here are some sample statements. Do yours sound like these?

1. In 1989 in the United States, men and women aged 35–44 earned more than men and women aged 25–34.

2. In 1989 in the United States, men of all ages earned more than women.

▶ EXERCISE 3

Use the graph on page 42 to decide if each statement is true or false.

T F **1.** The group with the highest average yearly earnings was men aged 45–54.

T F **2.** In 1989, at no time between the ages of 25 and 64 were women's average yearly earnings higher than the *lowest* average earnings for men between ages 25 and 64.

T F **3.** Men's average yearly earnings were at their highest at ages 55–64.

T F **4.** Women aged 25–34 earned a yearly average of $20,000.

▶ Answers are on page 178.

LINE GRAPHS

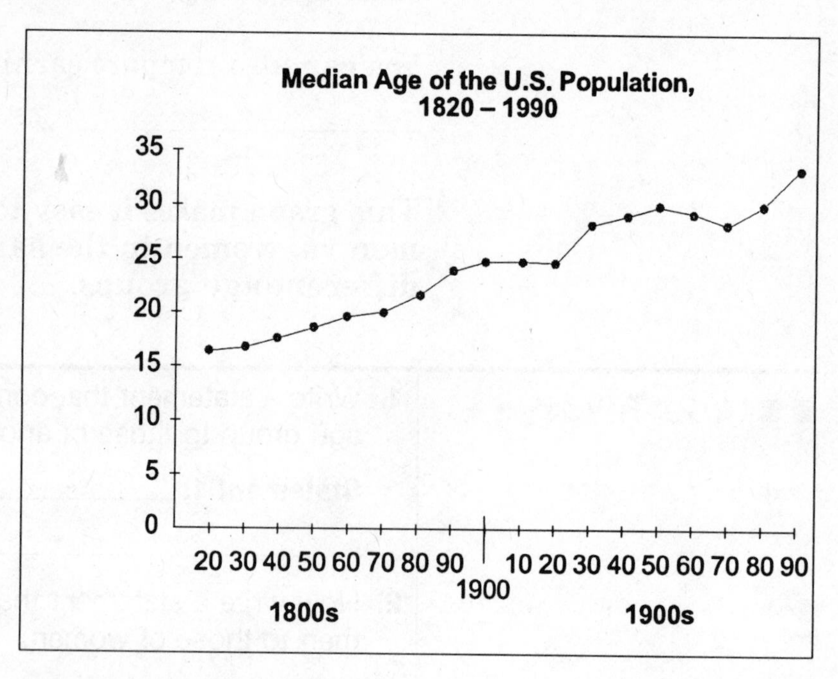

Source: U.S. Bureau of the Census

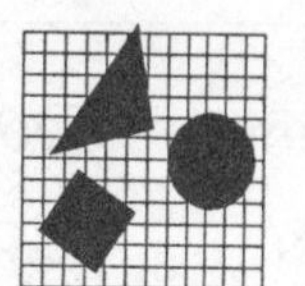

How is this line graph different from a bar graph?

Like a bar graph, a **line graph** uses values plotted along vertical and horizontal axes to display data visually.

You read a line graph in much the same way that you read a bar graph—by finding the point at which the vertical axis value and the horizontal axis value meet, or intersect. Instead of looking at a bar, however, you are looking at a line that may rise or fall, depending on the data.

YOU TRY IT ▶▶

What was the median age in the United States in 1920?

Step 1. Find the year 1920 along the horizontal axis.

Step 2. Find the point on the line graph directly above 1920.

Step 3. Now move your finger across and find the corresponding value on the vertical axis.

What is the vertical axis value?

The median U.S. age in 1920 was **25** years old.

A line graph shows values as they change over time. In the line graph on page 44, what value changes over time?

____

You're right if you said **median age in the United States**. In general, what changes took place between 1820 and 1990? Did the median age go up or down?

Just by looking at the angle of the line, you can see that **the median U.S. age went up**, except in the 1950s and 1960s, when it went down slightly.

● ● MATH RECAP ▶▶

Do you know what *median* means? In this case, it means that half the U.S. population is above the age on the graph, and half the population is below. You will learn more about median on page 62.

▶ EXERCISE 4

Part One

Use the graph on page 44 to answer the following questions.

1. What was the median U.S. age in 1950? __________
2. By how much did the median U.S. age go up between 1870 and 1920? __________
3. A woman is 35 in 1990. Is she above or below the median age in the United States? __________

Part Two

Make three statements using the graph on page 44. The first one has been started for you.

1. The median U.S. age was the same in 1980 as it was in ______________________________.
2. ______________________________
3. ______________________________

▶ Answers are on page 178.

CRITICAL THINKING

Based on your knowledge of history, why do you think the median U.S. age went down in the 1950s and 1960s? Were people dying at a younger age? Or was there a greater number of young people for some reason?

MORE LINE GRAPHS

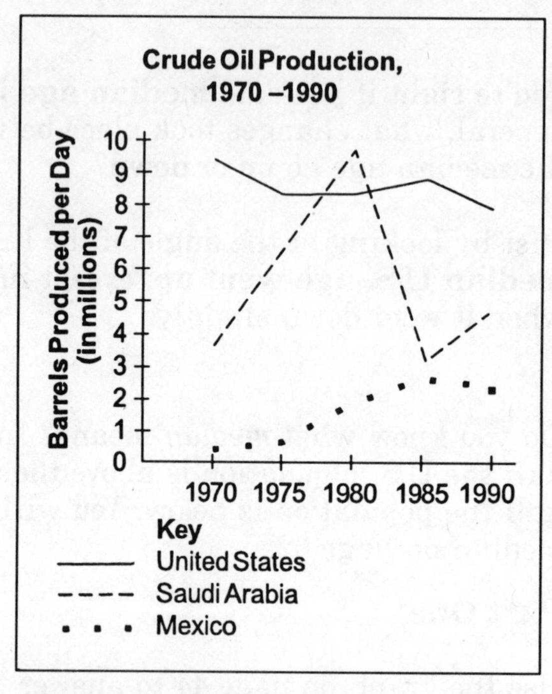

Source: U.S. Department of Energy
The Universal Almanac 1991, copyright 1990 by John Wright. Reprinted with permission of Andrews & McMeel.

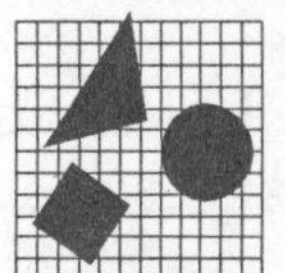

How is this line graph different from the line graph in the last lesson?

As in the line graph on page 44, this graph has years labeled on the horizontal axis, and it shows how certain values change over time. However, this line graph shows *three* separate lines, not just one. Each line represents a separate set of values. The key included on the graph shows what each line refers to.

For example, the solid line (——) refers to the United States. What do the other two lines refer to?

(• • •) refers to ______________________.

(- - -) refers to ______________________.

You're right if you said **Mexico** and **Saudi Arabia**, in that order.

As the title and axes of the graph indicate, all three lines show the changes in crude oil production between 1970 and 1990. Each individual line refers to a particular country.

YOU TRY IT ▶▶

Did the Saudi Arabian crude oil production rise or fall between 1980 and 1985?

Step 1. Find the line in the key that represents Saudi Arabia.	**1.** What does the line look like? ________
Step 2. Find the years 1980 and 1985 on the horizontal axis.	**2.** In 1980, Saudi Arabia produced about ________ million barrels per day.
Step 3. Find the points on the line graph directly above 1980 and 1985.	**3.** In 1985, Saudi Arabia produced about ________ million barrels per day.
Step 4. Now move your finger across and find the corresponding values for 1980 and 1985 on the vertical axis.	**4.** Saudi Arabian crude oil production ________ between 1980 and 1985.

STATS

Did you know that only 15% of homes in the United States in 1989 were heated by oil? Compare this figure to the 55% that were heated by natural gas.

▶ Answers are on page 178.

Notice that you could also answer the question above without looking at any specific values on the vertical axis. Because the data line is falling between 1980 and 1985, you know that Saudi crude oil production **fell** during that time.

▶ **EXERCISE 5**

Use the graph on page 46 to answer the questions.

1. In general, has Mexico's crude oil production risen or fallen between 1970 and 1990? ____________
2. Between 1985 and 1990, did the United States increase or decrease its crude oil production? ____________
3. Which country produced the most crude oil in 1980? ____________
4. Which country produced the most crude oil in 1985? ____________
5. Between 1980 and 1990, which country showed the greatest change in its crude oil production? ____________

▶ Answers are on page 178.

▶ EXERCISE 6

Part One

Use the data from the bar graph below to construct a line graph on a separate sheet of paper.

Step 1. Draw and label the vertical and horizontal axes using the same values that are given on the bar graph.

Step 2. Read the data for 1988 on the bar graph. On your line graph, put a dot or small × at the same value.

Step 3. Continue graphing data for the remaining years.

Step 4. When you are finished, connect the data points.

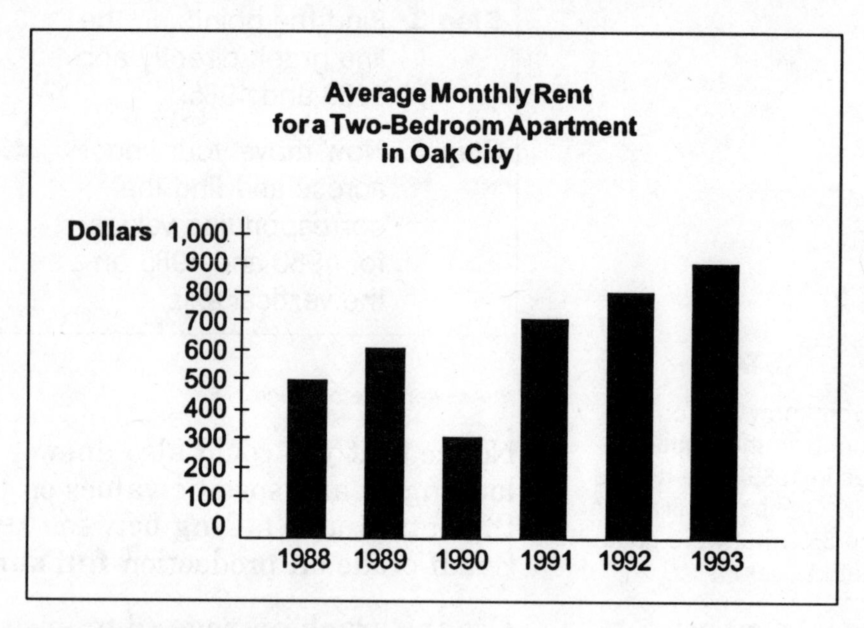

Source: Oak City Real Estate Commission

Part Two

1. Which graph (the bar graph above or the line graph you have just drawn) best shows the change in rental prices?

 ____

2. Write a statement comparing the highest average monthly rent with the lowest. ______________________________

 ____

3. Write a statement about Oak City rental prices between 1988 and 1993. ______________________________

 ____

▶ Answers are on page 178.

VISUAL STATISTICS

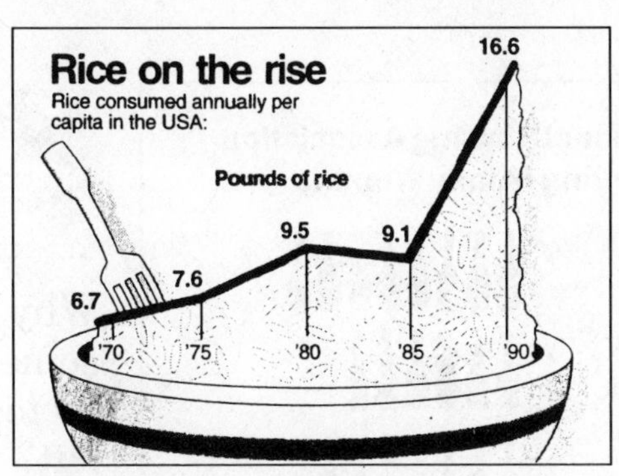

Source: Copyright 1991 by *USA TODAY*. Reprinted with permission.

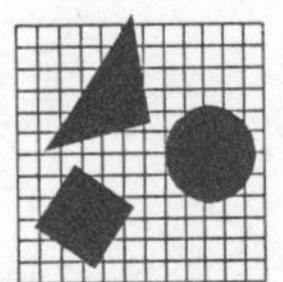

- Does this graph look like a line graph?
- How is it different from other line graphs you have worked with?

Although the graph uses a line to show data points, it is not really a line graph. In this case, no vertical axis is shown.

Per Capita: by each person

Some specific points are labeled to show the number of pounds of rice consumed **per capita** in 1970, 1975, 1980, 1985, and 1990. Instead of finding the exact values on a vertical axis, these points are labeled right on the graph. They're easy to read, aren't they?

Why don't all line graphs label data points like this? Here's why: Suppose you wanted to find the number of pounds of rice consumed per capita in 1986. It's not labeled on this graph, and there's no vertical axis with a range of values to provide additional information. On a real line graph, you could estimate the value for 1986 by using the scale on the vertical axis.

The visual statistics above are terrific if you want to know a general trend about rice consumption or if you are interested in the specific years labeled. Otherwise, a more complete line graph is better.

▶ **EXERCISE 7**

Use the visual statistics above to make two statements about rice consumption in the United States. The first one has been started for you.

1. Between 1970 and 1990, the annual per capita rice consumption in the United States ______________________

   ____.

2. ____

   ____

▶ Sample answers are on page 178.

PICTOGRAPHS

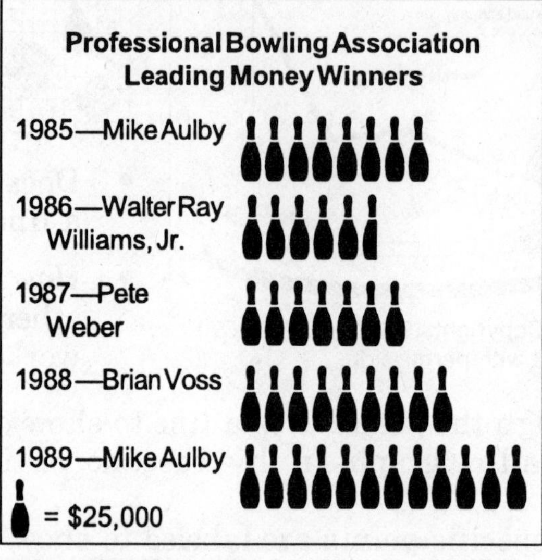

Source: *The Universal Almanac 1991*, copyright 1990 by John Wright. Reprinted with permission of Andrews & McMeel.

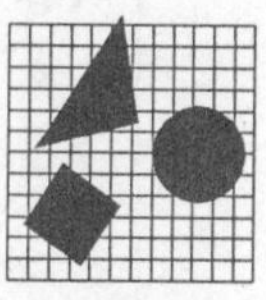

Why do you think data is sometimes shown on a pictograph like the one at left?

Pictographs are helpful for making comparisons at a glance. The pictograph above shows approximations of how much money top bowlers won.

The key tells you that each  symbol is equal to $25,000.

YOU TRY IT

How much money did Mike Aulby win in 1989?

Step 1. Find the correct year and name on the graph.

Step 2. Count the number of symbols next to his name. There are ________ symbols.

Step 3. Multiply that number by the value given in the key. ______ × ______ = $______

Did you find that Mike Aulby won **$300,000** in 1989?

EXERCISE 8

Now find and compare other values found in the graph.

1. How much more money did Mike Aulby win in 1989 than he did in 1985? ________________
2. Of the years shown, in which year did the leading money winner earn the smallest amount of money? ____________
3. What year was Brian Voss the leading money winner, and how much did he win? ________________

STATS

Did you know that the leading money winner in professional golf in 1989 was Tom Kite, who won $1,395,278? How does this compare to the leading money winner in bowling for that year?

4. For 1985 to 1989, did all of the leading money winners earn over $100,000? ______________

5. How much money did Walter Ray Williams, Jr., win in 1986? (Hint: $\frac{1}{2}$ of the symbol equals $\frac{1}{2}$ of $25,000.) ________

▶ Answers are on page 178.

Pictographs are often used in magazines and newspapers because they

- have visual appeal
- are easy to read
- can be easily constructed

▶ EXERCISE 9

Part One
Use the chart below to construct a pictograph.

- When deciding what dollar value your symbol should represent, look at the largest number you need to graph. Be sure that this amount can be represented in 15 or fewer symbols. More than 15 symbols can become difficult for your reader to count.
- Some suggestions for symbols are circles, squares, Xs, or dollar signs. Or you can create your own symbol.

Indianapolis 500
Approximate Total Prize Money

1985: $3,000,000
1986: $4,000,000
1987: $4,500,000
1988: $5,000,000
1989: $6,000,000
1990: $6,500,000

Part Two
Write two questions you could ask about the pictograph.

Example: Did the prize money stay the same two years in a row?

1. ____
____

2. ____
____

▶ Answers are on page 178.

CIRCLE GRAPHS

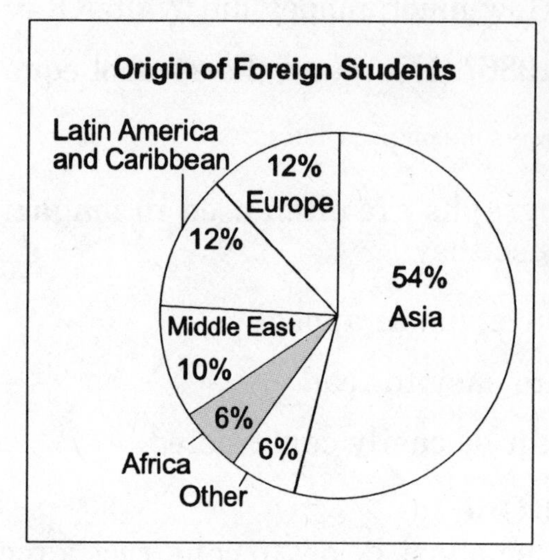

Source: Institute of International Education
Copyright 1991 by *USA Today*, Reprinted with permission.

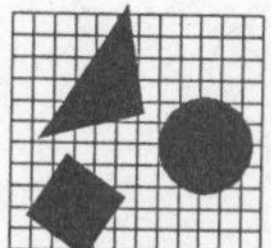

How do you think a circle graph differs from a bar or line graph?

A **circle graph** is just a circle divided into parts. The circle represents a *whole*, and each wedge, or segment, represents a *part* of that whole.

Add up the percents in the graph above. What total percent is represented by the circle graph?

The circle above represents *all*, or 100%, of the foreign students studying at U.S. colleges. Each part of that whole represents the percent from a specific area of the world.

YOU TRY IT

What percent of foreign students studying in the United States are from Africa?

- Find the segment, or part, labeled *Africa*.
- What percent is shown within this segment? ________%

You should have found that **6%** of foreign students studying in the United States are from Africa.

Circle graphs are very useful when you want to make comparisons. For example, just by looking at the graph above, you can see that a much larger percent of students come from Asia than from Africa or the Middle East.

■■ **YOU TRY IT** ▶▶

Make two statements comparing the percent of students from different regions. The first one has been started for you.

1. There are fewer students from ______________________________ ______________________________.

2. ______________________________ ______________________________

▶ Sample statements are on page 178.

What does *Other* refer to on the graph? There are often groups of data that are too small to list. These are combined to form one larger segment labeled *Other*. In this way, the graph still represents 100% of the data.

"Dollar" Circle Graphs

Circle graphs are usually divided into percents, as is the one on page 52. Sometimes, however, a circle graph is divided into *cents*, or parts of a dollar. Look at the graph below.

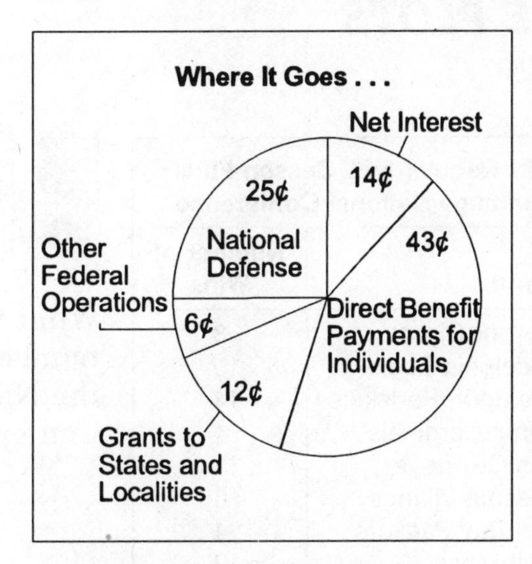

Source: Office of Management and Budget
The Universal Almanac 1991, copyright 1990 by John Wright. Reprinted with permission from Andrews and McMeel. All rights reserved.

The graph at left shows you how many cents out of every dollar went toward various government functions in 1991.

For example, out of every dollar collected by the federal government, 25 cents went toward national defense. How many cents went toward grants to states and localities?

You're correct if you said **12¢**.

■■ **YOU TRY IT** ▶▶

Make a statement comparing the amount of federal money that goes to two different government functions. ______________________________ ______________________________

▶ A sample statement is on page 179.

▶ EXERCISE 10 Use the information below to label the circle graph. The circle represents a total of 100¢ and the segments are proportional. For example, 43¢ is a little less than $\frac{1}{2}$ of the circle.

. . . The United States Office of Management and Budget reports that 34 cents out of every dollar collected by the U.S. government comes from Social Security receipts and 43 cents out of every dollar comes from individual income taxes. Out of each dollar, it is calculated that 11 cents comes from corporation income tax. The remaining 12 cents comes from other sources. . . .

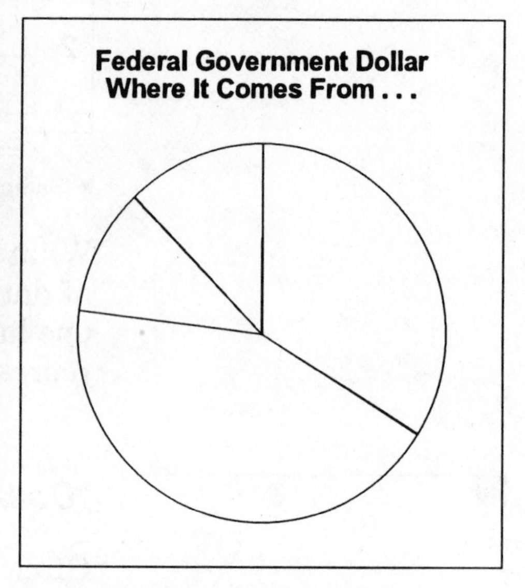

Source: Office of Management and Budget

▶ Answers are on page 179.

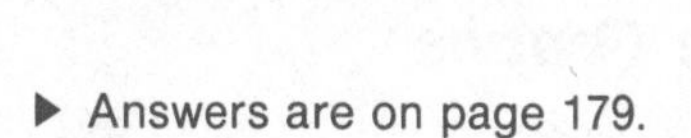

LINE PLOTS

1991 Regular NFL Season Final Standings National Conference	
Team	**Number of Wins**
New York Giants	8
Philadelphia Eagles	10
Washington Redskins	14
Phoenix Cardinals	4
Dallas Cowboys	11
Minnesota Vikings	8
Green Bay Packers	4
Detroit Lions	12
Chicago Bears	11
Tampa Bay Buccaneers	3
San Francisco 49ers	10
Los Angeles Rams	3
New Orleans Saints	11
Atlanta Falcons	10

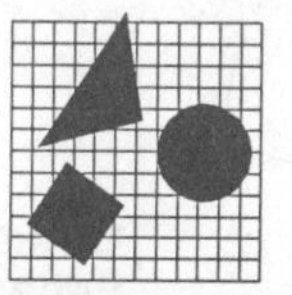

What was the most common number of wins by teams in the National Football Conference in 1991?

A **line plot** is often used to show data in a more visual way than the chart on page 54. Let's construct a line plot and see what advantages it has.

■■ You Try It ▶▶

Step 1. Draw a line across your paper.

Step 2. Find the scale you are working with. (In this case, the fewest number of wins is 3, and the largest is 14.) Write these numbers under your line as shown and fill in the values in between.

3 4 5 6 7 8 9 10 11 12 13 14

Step 3. Now use an × to record each number of wins on the number line. (The Giants had 8 wins, so put an × above the 8 on the number line.)

x

3 4 5 6 7 8 9 10 11 12 13 14

Step 4. Continue marking the ×s in a stack above the appropriate numbers. (The values for the Eagles, the Redskins, the Cardinals, the Cowboys, and the Vikings are now added.)

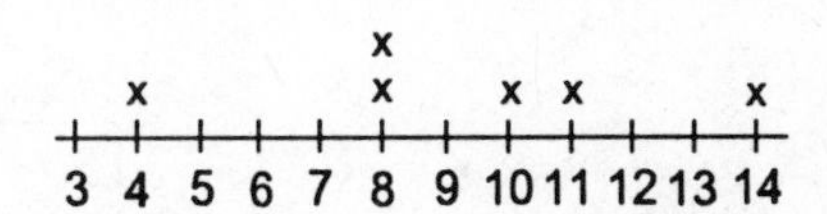

▶ Answers are on page 179.

What types of things can you see more clearly from a line plot than you can from a chart of data?

You may have seen a couple of things just "jump out at you" that perhaps were not as obvious when you read the chart.

Cluster: an area on a line plot where more values are recorded than elsewhere on the line

1. There is a **cluster** of ×s around a certain area of the line. Where is this bunch, or group, of ×s? You're right if you said **around 10 and 11**. This cluster of ×s tells you that 10 or 11 wins was a more common occurrence than, say, 6 or 7 wins.

2. There are single data points on the line that are quite a bit above or below the cluster. From these, it is also easy to see the highest and lowest values in the data collected.

Line plots make it easy to compare individual values at a glance. For instance, do teams generally win about 5 games, or more than that?

▶ EXERCISE 11 Part One

Use the data below to create your own line plot.

- Read the number of representatives for each state and put an × above that value on the line plot.
- The first five states on the list have been marked for you.

Number of Members in the House of Representatives, by State, 1990

State	Members	State	Members	State	Members
✓ Alabama	7	Louisiana	8	Ohio	21
✓ Alaska	2	Maine	2	Oklahoma	6
✓ Arizona	5	Maryland	8	Oregon	5
✓ Arkansas	4	Massachusetts	11	Pennsylvania	23
✓ California	45	Michigan	18	Rhode Island	2
Colorado	6	Minnesota	8	South Carolina	6
Connecticut	6	Mississippi	5	South Dakota	2
Delaware	2	Missouri	9	Tennessee	9
Florida	19	Montana	2	Texas	27
Georgia	10	Nebraska	3	Utah	3
Hawaii	2	Nevada	2	Vermont	4
Idaho	2	New Hampshire	2	Virginia	10
Illinois	22	New Jersey	14	Washington	8
Indiana	10	New Mexico	3	West Virginia	9
Iowa	6	New York	34	Wisconsin	9
Kansas	5	North Carolina	11	Wyoming	2
Kentucky	7	North Dakota	2		

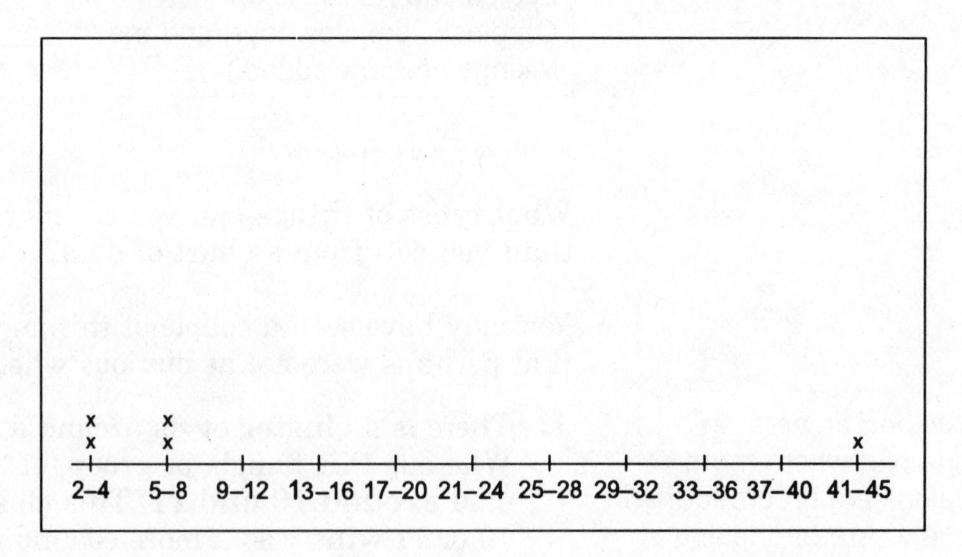

Part Two

1. What is the smallest number of representatives from a state, and how many states have that many?

____

____

2. What is the largest number of representatives? Are there many states that have close to this number of representatives?

____

____

3. Is there a cluster of values in a certain area of the line plot? Where is it?

____

____

4. Representation in the House is based on state population. The larger the state population, the greater the number of representatives. Make two statements about population based on the information above. The first one has been started for you.

   a. Many states have _______________

   _______________.

   b. _______________

   ____

▶ Answers are on page 179.

CRITICAL THINKING

Each state in the United States is represented by two members in the Senate, regardless of the state's population. Why do you think membership in the House of Representatives is based on population? Is it fair that states with more people have more representatives?

SCATTER DIAGRAMS

An employee of a cereal manufacturer said to his boss, "I have this theory that the more sugar we put in our breakfast cereals, the more kids will like them. What do you think?" "Brilliant!" replied the boss. "Let's collect some data and check it out." Here's the data they came up with:

Cereal	Teaspoons of Sugar per Serving	Number of Children Who Liked It
Good for You	1	2
Sweeties	8	19
YumYums	4	12
SugarPlus	5	11
PlainPuffs	2	5
Sugar City	9	20
Kiddo Krunch	5	13
Sweethearts	8	17
Munchies	6	12

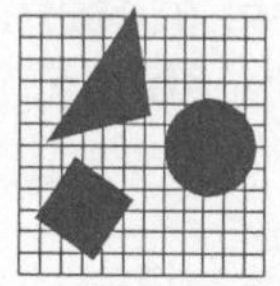

- Is the employee's theory correct?
- Can you tell by looking at the chart?

You may have noticed that it is difficult to prove the employee's theory by simply looking at the data. An excellent way to display this type of data is with a **scatter diagram.** The diagram below shows the intersection of amount of sugar and number of children.

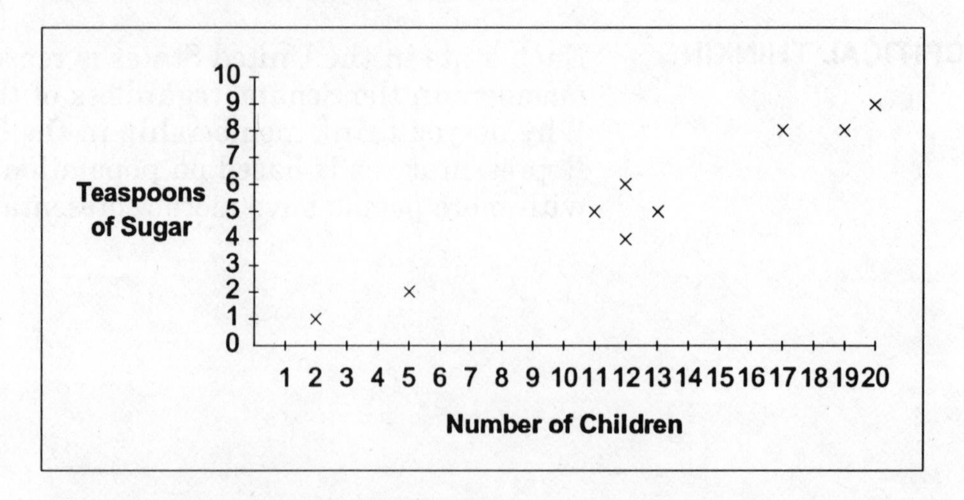

Can you tell from this diagram whether the cereal company employee's theory is true? Does more sugar mean that more kids like the cereal?

Correlation: relationship that shows a trend and allows you to draw conclusions

The answer is **yes**. As the number of teaspoons of sugar goes up, so does the number of kids who like the cereal. In general, as one number gets larger, the other number gets larger. This means that there is a **correlation** between the two values.

▶ EXERCISE 12

A research group did an informal study of the relationship between people's income and the number of years of school they had completed.

Some of the responses have already been plotted on the scatter diagram below.

Now you plot the responses listed in the chart at right. For each respondent, put an × at the intersection of the two values: years of school and income.

Respondent	Years of School	Income
#2E	16	$30,000
2F	10	28,000
2G	14	21,000
3A	14	20,000
3B	8	7,000
3C	9	10,000
3D	9	11,000
3E	15	30,000
3F	16	25,000
3G	16	12,000

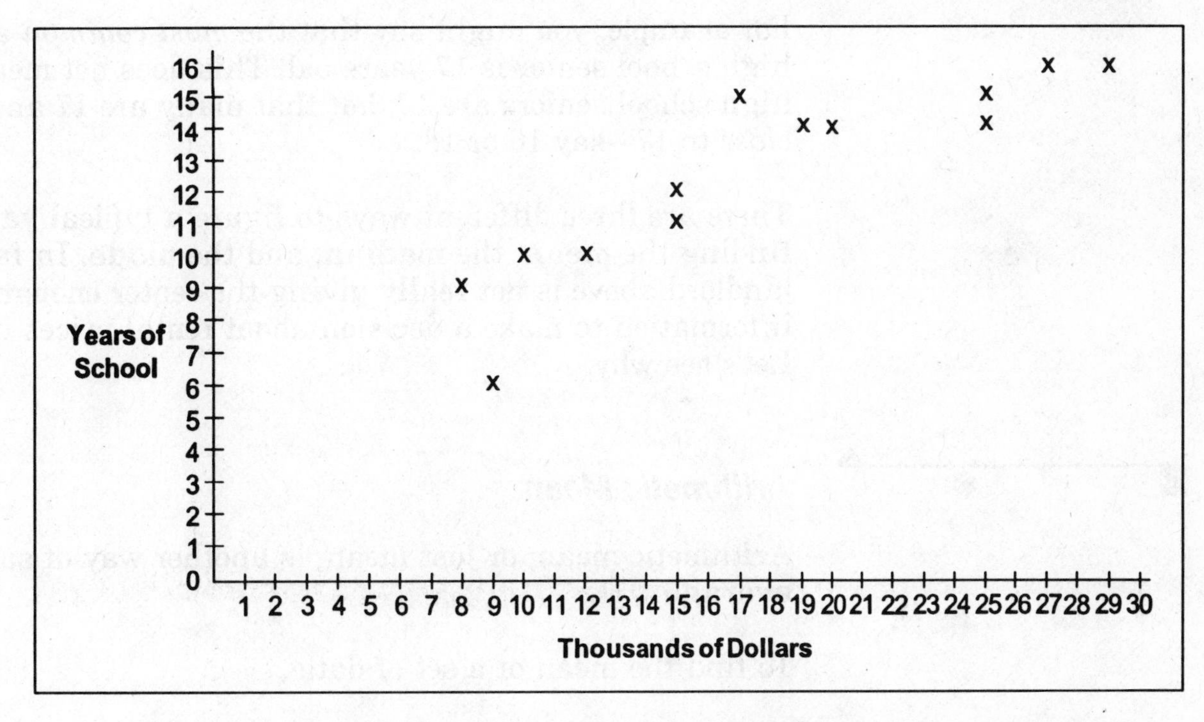

1. In general, do you see a connection between number of school years and income? ____________

2. Write a statement about the data based on your scatter diagram. ____________

____

▶ Answers are on page 179.

WHAT IS A TYPICAL VALUE?

"You are getting a fantastic deal on this apartment," says a landlord. "This beautiful place for only $400 per month! The average rental rate for two-bedroom apartments in this neighborhood is almost $500. If you want to live in this area, you can't do better than this place."

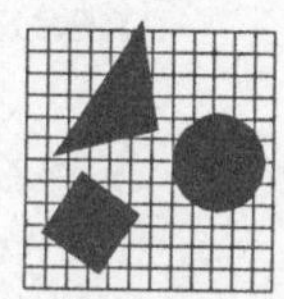

- Sounds like a good deal, doesn't it?
- Assuming that the landlord did the math correctly, do you think you could get a two-bedroom apartment for much less than $400?

You have seen how data and statistics can be represented on graphs, charts, and diagrams. Another way that statistics are often presented is in the form of **typical values**.

For example, you might say that the *most common* age of a high school senior is 17 years old. This does not mean that *all* high school seniors are 17, but that many are 17 and most are *close* to 17—say 16 or 18.

There are three different ways to figure a typical value: finding the **mean**, the **median**, and the **mode**. In fact, the landlord above is not really giving the renter enough information to make a decision about rental prices in the area. Let's see why.

Arithmetic Mean

Arithmetic mean, or just mean, is another way of saying **average**.

To find the mean of a set of data

Step 1. Add up the numbers.

Step 2. Divide the sum by the number of numbers in your set of data.

Say, for example, that the rental rates in the neighborhood referred to on page 60 are as follows:

1200 Speen St.	$250	1320 Day St.	$240
1202 Speen St.	250	1322 Day St.	220
1203 Speen St.	240	1400 Day St.	250
1211 Speen St.	250	1401 Day St.	250
1220 Speen St.	250	400 North Ave.	250
320 Ashland Ave.	270	402 North Ave.	290
330 Ashland Ave.	240	404 North Ave.	270
333 Ashland Ave.	260	406 North Ave.	260
440 Ashland Ave.	230	408 North Ave.	1,700
500 Ashland Ave.	250	410 North Ave.	1,650
1111 Day St.	290	412 North Ave.	1,600
1290 Day St.	250	414 North Ave.	1,700

YOU TRY IT ▶▶

1. Add up the numbers in the set of data. — The total is **$11,710.**
2. Count the number of rental rates listed above. — There are ______ numbers in this set of data.
3. Divide by the number of numbers in the set of data. — ______ ÷ ______ = ______ The mean is $______

▶ Answers are on page 179.

As the landlord stated, the average (mean) of rental rates is "almost $500." However, look again at the rates listed above. Are any of the values even close to $500? No. In fact, most of the values *center around $250.*

The mean is so high because of the four very high rental rates of $1,700, $1,600, $1,700, and $1,650. These rates are so much higher than the others that they raise the mean to an amount that is not really typical!

Is the landlord on page 60 not telling the truth? Is it possible to tell the truth but still be misleading?

▶ EXERCISE 13 Find the mean (average) in each set of data below.

1. Distance from home to workplace among employees: 2.8 miles, 4.8 miles, 8 miles, 1.5 miles, 3 miles, 6 miles, 2.5 miles, 3.4 miles.
2. Weekly grocery bills: $90, $120, $48, $55, $37.
3. Tips earned: $35, $48, $22, $18, $25, $32, $44.

▶ Answers are on page 179.

MEDIAN

Let's look at a second way to describe the data of rental rates from page 61.

The **median** of a set of data is the middle value when the pieces of data are ordered from smallest to largest.

For example, what is the median in this data set?

56	61	49	32	90	75	23	101	70

Step 1. Put the set of data in order, from smallest to largest.

23	32	49	56	61	70	75	90	101

Step 2. Select the value in the middle of the set.

23	32	49	56	(61)	70	75	90	101

61 is the median value in this set of data.

Note: When there is an even number of values in a set of data, the median is the *mean* of the middle *two* numbers.

Example: 20, 21, (22, 23,) 24, 25

$22 + 23 = 45 \qquad 45 \div 2 = 22\frac{1}{2}$

mean of 22 and 23

So, $22\frac{1}{2}$ is the median of the set.

Looking back to the rental rates given on page 61, the **median** rental rate is $250.

Can you see that the **median** value of $250 is much more representative of the rental rates in this neighborhood than the mean value of $487?

▶ EXERCISE 14

Find the median in each set of data below.

1. Calculator prices among area stores: $9.50, $12, $14.50, $12, $12.50, $13.
2. Height of South High basketball players: 6′2″, 6′10″, 7′3″, 7′4″, 5′11″.

▶ Answers are on page 179.

MODE

Now let's look at the third way to find a typical value. The **mode** in a set of data is the number that occurs most frequently in the set.

For example, what is the mode in this set of data?

21	15	70	22	21	71	21	15	40

How many times does each number appear?

21: ____ 70: ____ 71: ____

15: ____ 22: ____ 40: ____

21 is the mode, since it appears most often in the set of data.

YOU TRY IT

What is the mode of the rental rates given on page 61?

1. Determine how many times each value appears in the set of data. (This process has been started for you.)

1,700: 2 260: ____
1,650: ____ 250: ____
1,600: ____ 240: ____
290: ____ 230: ____
270: ____ 220: ____

2. Which value appears most frequently? The mode is ____.

▶ Answers are on page 179.

Again, in this case, the **mode** gives you a better picture of the rental rates in the area than the mean does.

Note: Sometimes there is no mode in a set of data because each value appears only once. In this case, use mean or median to express the typical value.

EXERCISE 15 Find the mode in each set of data below.

1. Prices of a 1-lb. box of salt in area stores: $.57, $.49, $.52, $.49, $.57, $.49.
2. Number of weeks on Top 40 radio: 8, 10, 5, 3, 9, 10, 10, 9, 5, 8, 4, 2.

▶ Answers are on page 180.

MEAN, MEDIAN, AND MODE

Is mode or median always a better way to show the typical value than the mean? No. In the case of the landlord on page 60, it certainly would have been fairer to the potential renter to use median or mode. However, there are other cases where the mean *does* give a more accurate picture of the typical value. A very general guideline follows:

> Use the mean to show typical value *unless* there are extreme values in the set of data. Extreme values are those that are not typical of the rest of the data.

▶ **EXERCISE 16** For each set of data below, find the mean, the median, and the mode. Then decide which method best represents the most typical value in each set.

1. Dollars Collected by Charity Groups

Cancer Fund	$1,100	Mean: ________
Grand Foundation	300	
Women for Women	1,100	Median: ________
John Kramer Fund	650	
Heart & Lung Assn.	500	Mode: ________
Dalmore Committee	230	

Best representation of typical value: ________________

STATS

Did you know that . . . the *median* family income in the United States in 1991 was $32,191? Do you think the mean would be higher or lower? Why?

2. Attendance at Plainville Pirates Baseball Games

June 15 game	230 people	Mean: ________
June 16 game	175 people	
June 30 game	225 people	Median: ________
July 4 game	700 people	
July 5 game	225 people	Mode: ________
July 12 game	185 people	
July 14 game	200 people	
July 20 game	180 people	
July 27 game	185 people	

Best representation of typical value: ________

▶ Answers are on page 180.

CRITICAL THINKING WITH DATA

Suppose a local school system is deciding how much money to allot to its high school sports program. Part of its decision is based on how much the surrounding school systems spend on their programs. Use the graph below and your knowledge of typical values to answer the questions that follow.

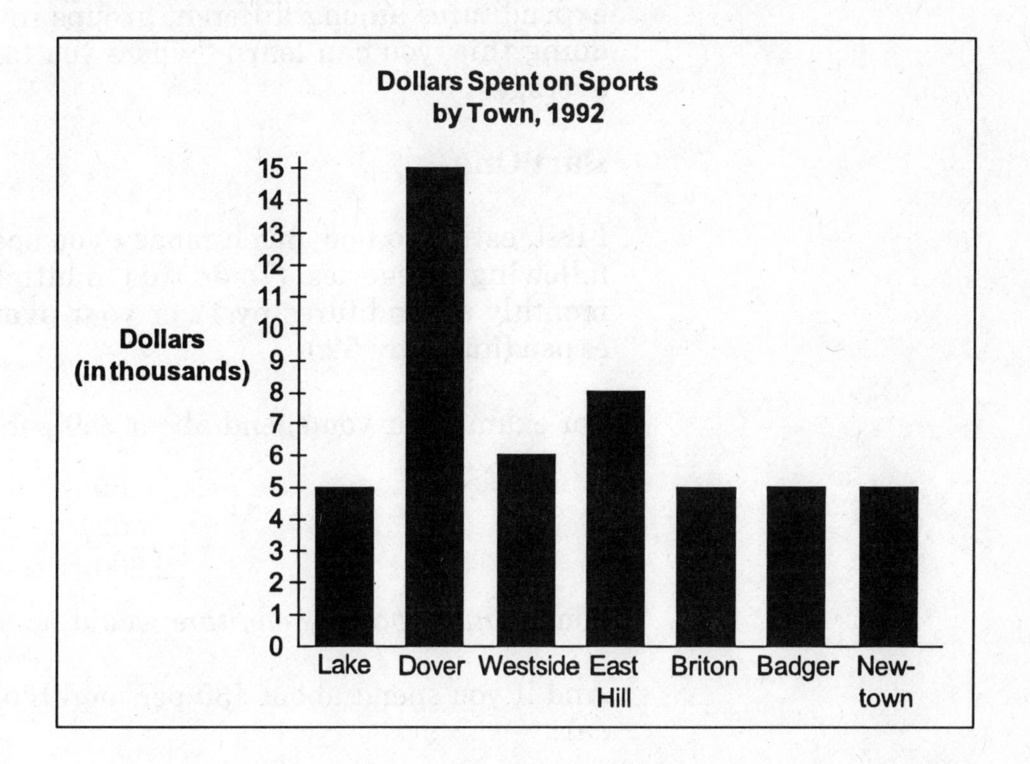

1. *Suppose you are a high school football coach in the Westside school system.* You want your school system to spend as much money as possible on the sports program. Would you use mean, median, or mode to show that Westside spends too little? Write a paragraph persuading the school system to spend more money. Use any data you need from the graph above.
2. *Now imagine that you are a Westside school department member, and you are tired of having so much money spent on sports.* You would rather have more money allotted to teachers' salaries. Would you use mean, median, or mode to show that Westside spends too much on sports? Write a paragraph defending your point of view. Use any data you need from the graph above.

Are You Average?

As you know, comparing data is a good way to learn more about a subject. In this lesson, you will compare the amount of money you spend on certain items to the **average** expenditures among different groups in the United States. By doing this, you can learn "where you fall" among national averages.

Part One

First, estimate how much money you spend *per year* in the following categories. (To do this, multiply your **average** monthly expenditures by 12 or your **average** weekly expenditures by 52.)

For example, if you spend about $50 per week on food:

$$\begin{array}{r} 50 \\ \times\ 52 \\ \hline 2{,}600 \end{array}$$

Your *annual food expenditure* would be $2,600.

And if you spend about $30 per month on gas and oil for your car:

$$\begin{array}{r} 30 \\ \times\ 12 \\ \hline 360 \end{array}$$

Your *annual gas/oil expenditure* would be $360.

Category	Annual Expenditure
Food, prepared at home	$ ________ per year
Food, away from home	$ ________ per year
Alcoholic beverages	$ ________ per year
Clothing	$ ________ per year
Gas and motor oil	$ ________ per year

Part Two

The chart below shows average expenditures in the United States by age, geographic region, and consumer unit. Circle the group that you belong in for each of the three categories: *age*, *region*, and *consumer unit*. Then answer the questions that follow.

Category	Food, home	Food, away	Alcoholic beverages	Clothing	Gas and motor oil
Age					
Under 25	$1,121	$1,334	$312	$1,042	$ 659
25–34	2,046	1,618	355	1,504	923
35–44	2,599	2,037	281	2,015	1,152
45–54	2,605	2,210	307	2,112	1,248
55–64	2,355	1,597	239	1,355	970
65–74	1,933	1,081	162	977	729
75 and over	1,373	566	89	451	367
Region					
Northeast	$2,212	$1,781	$272	$1,619	$ 812
Midwest	2,084	1,577	273	1,465	893
South	2,039	1,514	229	1,406	1,025
West	2,282	1,643	326	1,521	952
Consumer Unit					
Buying for one person	$ 989	$1,099	$278	$ 893	$ 518
Buying for two or more people	2,585	1,813	265	1,722	1,093

Source: U.S. Bureau of the Census

1. Among people in your *age group*, do you spend more or less than the average on *food away from home*? ________

2. Among people in your *geographic region*, do you spend more or less than the average on *gas and oil*? ________

3. Write a statement comparing your *clothing* expenditures to the average in your *age group*. ________

____

4. Find a category in which your expenditures are far above or below the average in your age, region, or consumer unit size. Why do you think you are not close to the average? Write a short explanation. ________

____

____

GROUP PROJECT

Organizing Data

Remember all those questionnaires you filled out back in Chapter One? Now it's time to make some sense of all that information. Here's what to do:

- If you are working in a group, add the numbers you came up with on page 31 to the numbers obtained by the rest of your group.
- Use the information on pages 66 and 67 to plot your data.

Part One
Chart/Table

1. Divide your questionnaires into a Male and Female group.
2. Divide each of these two groups into a "Day Person" group and a "Night Owl" group. (You should have four groups.) Count up each group and record the numbers below.
3. Put all questionnaires back into Male/Female groupings.
4. Divide them once again, this time into a Lives Alone group and a Lives with Others group. (You should have four new groups.) Count up each group and record the numbers below.
5. Make a statement about the data you have organized.

	Male	Female
"Day Person"		
"Night Owl"		
Lives Alone		
Lives with Others		

Statement: ______________________________

____

____

Part Two
Line Plot

From page 31, use the information you collected in answer to "How many hours of television do you watch per week?" to complete the line plot below.

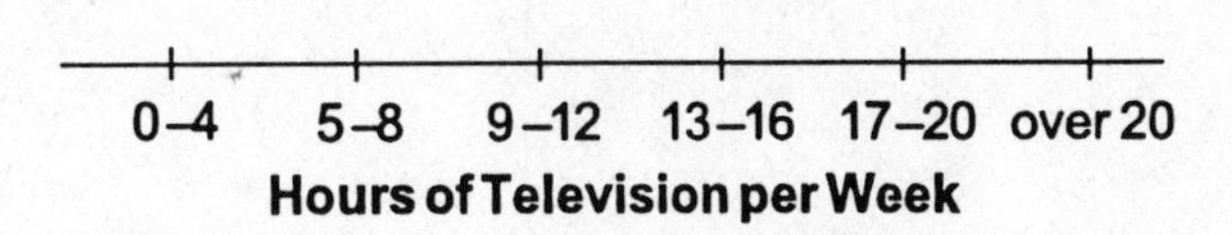

Statement: ______________________________

Part Three
Scatter Diagram

1. From page 31, use the data from:

   "How many times per week do you see people for a social occasion?"

   "How many close friends do you have, not including family?"

2. Label one axis below "Number of Social Occasions" and the other "Number of Close Friends."

3. Label the scale for each axis. Be sure to take into account the range of values for each.

4. Plot the data.

Statement: ______________________________

CHAPTER THREE

Analyzing Data

THE PURPOSE OF A GRAPH OR CHART

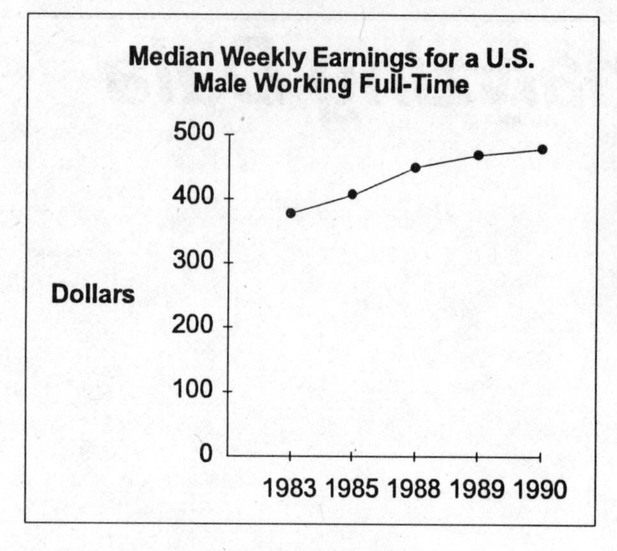

Source: U.S. Bureau of Labor Statistics

What is the purpose of the graph at left?

Your first step in analyzing data is to understand the *topic* of the data. Many times, mistakes are made in interpreting data because a reader did not understand the goal, or purpose, of a graph or chart. For example, put a check mark next to the statement that accurately summarizes the purpose of the graph.

For the period 1983 to 1990, the graph shows the rise in

☐ **a.** median weekly earnings in the United States

☐ **b.** median weekly earnings for a U.S. male working full-time

☐ **c.** median weekly earnings for a U.S. male

You're correct if you chose statement **b**. What's wrong with the other two?

1. If the first statement were true, you could expect to find information about *female* workers. The graph title clearly states *male* workers.
2. If the third statement were true, you could expect to find information about *part-time* workers' earnings. The graph title specifies *full-time.*

Only statement **b** accurately summarizes the graph's purpose.

To understand the purpose of a chart, graph, or table, you need to use information from the *title* and *labels*. Ask yourself these basic questions:

- **What** data is being presented?
- **When**, or over what period of time, does the data apply?
- **Where**, or what location, is the data concerned with?
- **How** is the data presented (in dollars? in miles per hour? in average cost?)

YOU TRY IT

Fill in the blanks to summarize the purpose of the chart below.

Highest-Paying Jobs* in the United States, 1990

Profession	Average Weekly Salary
Lawyer	$990
Airline Pilot	807
Chemical Engineer	807
Electrical Engineer	803
Aerospace Engineer	801
Physician	792
Mechanical Engineer	766

*Among fields employing at least 50,000 people

Source: *Occupational Outlook Quarterly*

The chart shows the ______________ (what?)

in the ______________ (where?)

in ______________ (when?)

in ______________ (how?)

(among fields employing at least 50,000 people).

An accurate summary of the purpose for the chart above is

The chart shows the *highest-paying job* (what?)

in *the United States* (where?)

in *1990* (when?) in *average weekly salary* (how?)

(among fields employing at least 50,000 people).

Of course, there are other ways to summarize a graph or chart. Using your own words, state the purpose of the graph on page 74. Be sure to include the "what, when, where, and how" information in some form.

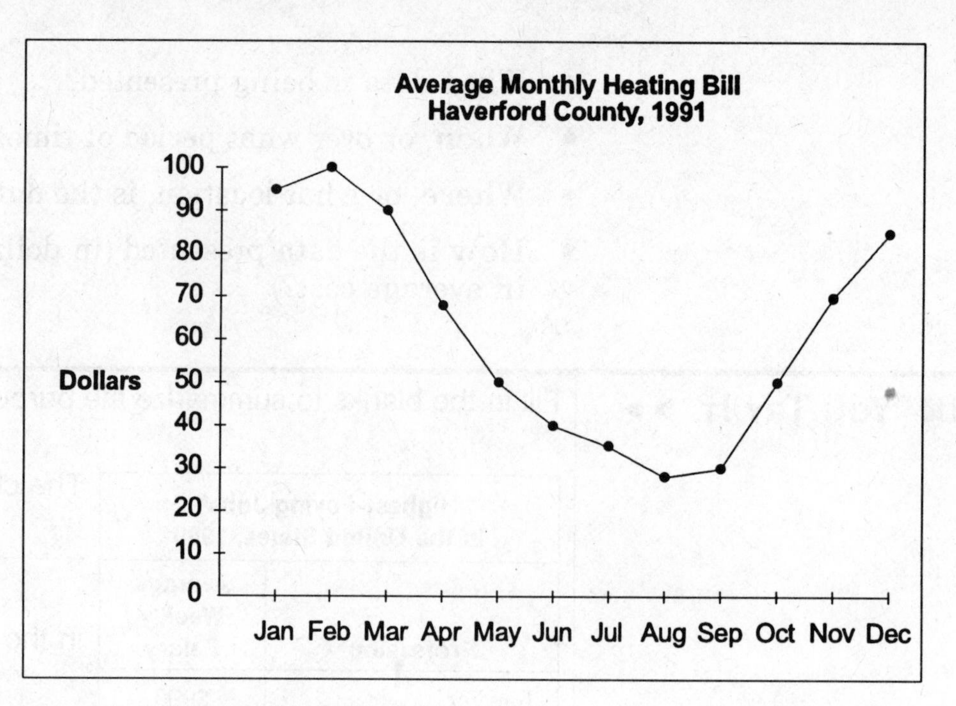

Source: County Utilities Commission

The graph shows the ______________________________
what?

in ______________________________ in ______________
where? when?

in ______________.
how?

Does your summary sound something like this?
The graph shows the average monthly heating bill in Haverford County in 1991 in dollars.

The Source of Data

All good graphs, charts, and tables name the **source** that collected and organized the data. A **source line** is usually printed somewhere beneath the data.

What is the source of the graph above showing average monthly heating bills in Haverford County?

You're correct if you said **County Utilities Commission**.

The source of any data is important to know in case you have questions about the data or its reliability.

▶ EXERCISE 1 Decide if the summary given for each set of data below is accurate. If it is not, which information is inaccurate—the what, where, when, or how?

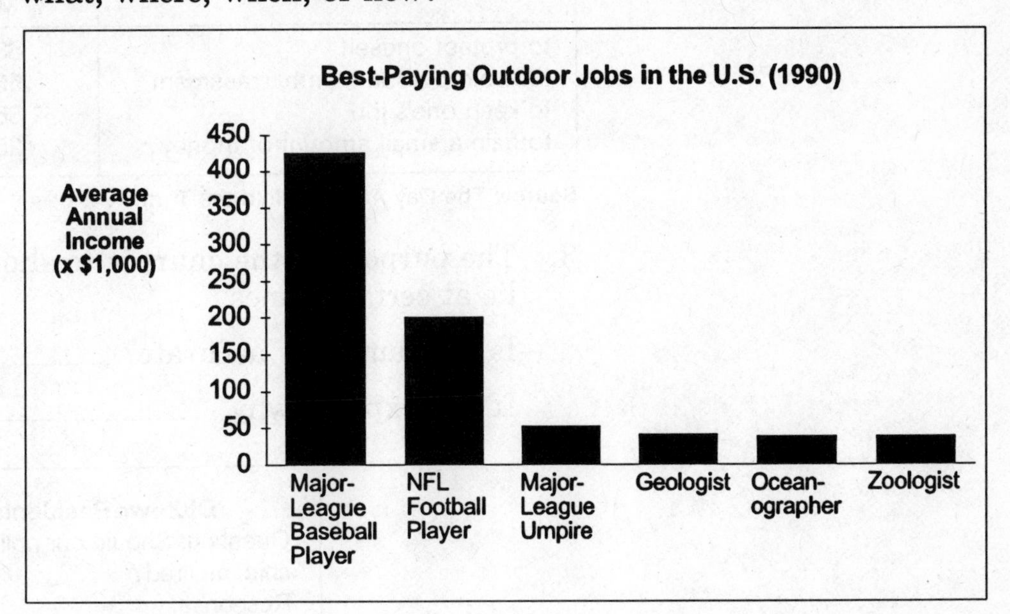

Source: *Jobs Rated Almanac*

1. The purpose of the graph is to show the 1990 average annual income for the best-paying jobs in the United States, in dollars.

   Is the summary accurate? _____

   If not, explain why. ______________________________

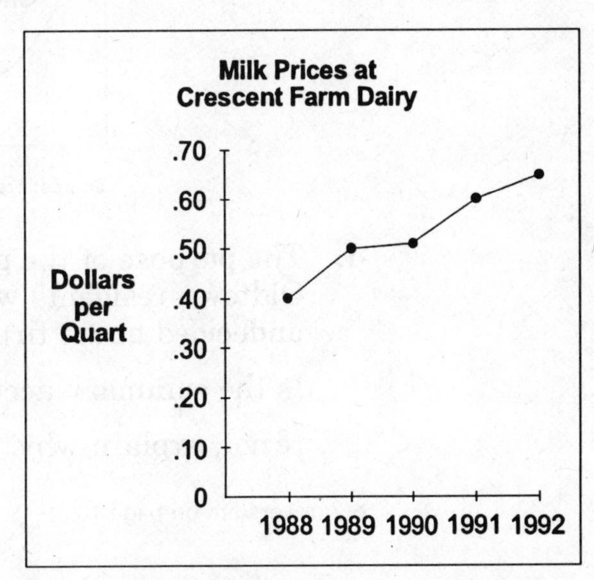

Source: County Dairy Association

2. The purpose of the graph is to show the price of milk from 1988 to 1992 in dollars per quart.

   Is the summary accurate? _____

   If not, explain why. ______________________________

Americans Who Believe It's All Right to Lie		
	% Men	% Women
to protect oneself	63	52
to avoid personal embarrassment	56	48
to keep one's job	56	35
to gain a small amount of money	25	15

Source: The Day America Told the Truth

3. The purpose of the chart is to show how many Americans lie at certain times.

Is the summary accurate? _____

If not, explain why. ______________________________

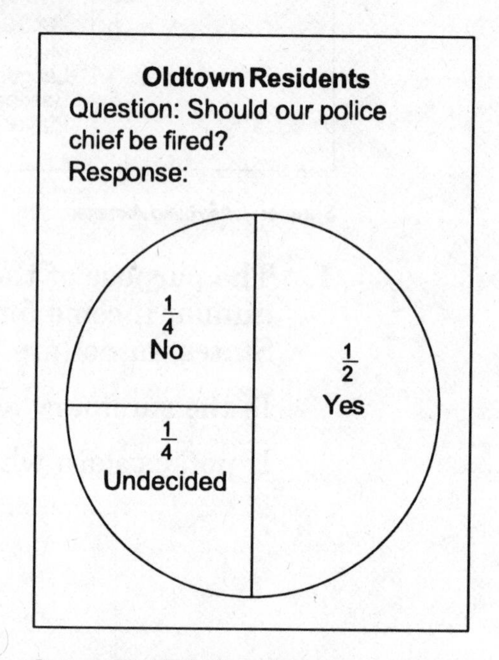

Source: *Free Daily Press* Survey

4. The purpose of the graph is to show what fraction of Oldtown residents were in favor of, not in favor of, or undecided about firing the police chief.

Is the summary accurate? _____

If not, explain why. ______________________________

▶ Answers are on page 180.

READING DATA: BE SPECIFIC

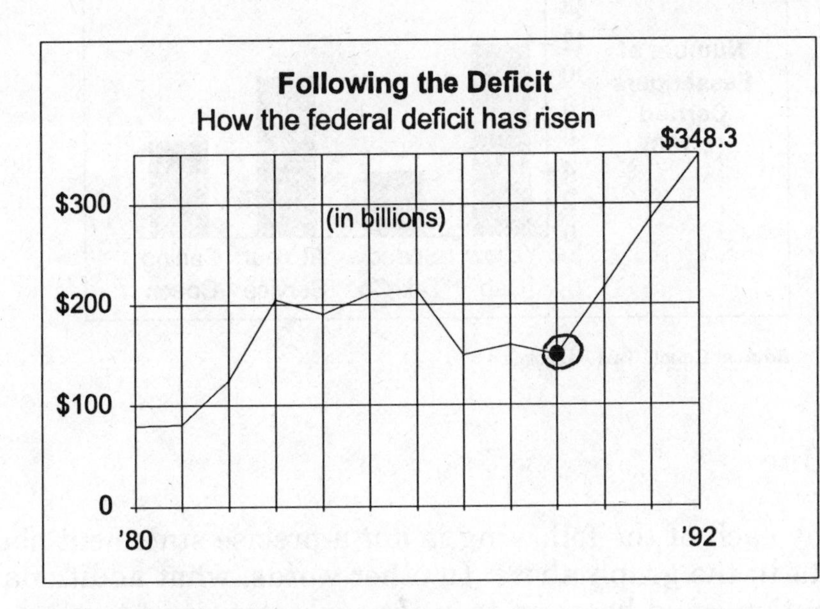

Source: U.S. Office of Management and Budget
Copyright 1991 by *USA Today*. Reprinted with permission.

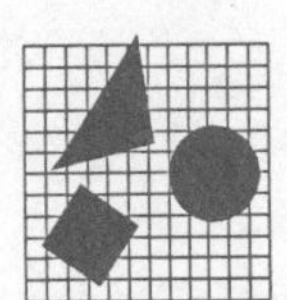

The vertical axis value of the data point circled is between $100 and $200. The horizontal axis value is between '80 and '92. But does the graph tell you more than this about the circled data point?

In order to be as accurate as possible when making statements about specific data points, you need to use *all* of the information provided, including the labels of *both* axes.

Look at the circled data point on the graph. Put a check mark next to the *most accurate* and *specific* statement about that data point.

☐ The federal deficit in 1989 was $150.

☐ Between 1980 and 1992, the federal deficit was $150 billion.

☐ In 1989, the federal deficit was $150 billion.

Now let's look at each statement in detail.

1. **The federal deficit in 1989 was $150.** *False.*

   This statement fails to take into account the label *in billions*. The correct figure is *$150 billion*.

2. **Between 1980 and 1992, the federal deficit was $150 billion.** *Not specific.*

   Not all of the years between 1980 and 1992 showed a deficit of $150 billion.

3. **In 1989, the federal deficit was $150 billion.** *Accurate.*

▶ EXERCISE 2

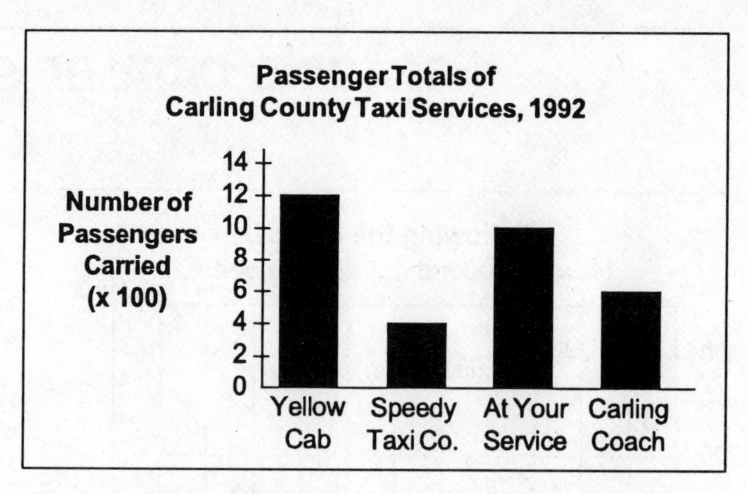

Source: County Taxi Association

Part One

Tell why each of the following is *not* a precise statement about the data in the graph above. In other words, what additional information could be given to make each statement true or more accurate?

Example: In 1992, the Yellow Cab Company carried 800 times as many passengers as Speedy Taxi Co. did.

The Yellow Cab Company carried 800 more passengers, not 800 times as many.

1. In 1992, the Yellow Cab Company in Carling County carried 12 passengers.

____

____

2. At Your Service carried 1,000 passengers.

____

____

3. The Carling Coach bar goes up to 600 on the graph.

____

____

Part Two

Using the graph on page 78, choose four phrases from the list below to make four different statements about the data. Be sure to use

- information from the title of the graph
- information from *both* axes
- labels

Phrases

$\frac{1}{3}$ the number of passengers	more passengers than
twice as many	800 more passengers than
half as many	200 fewer passengers than
the greatest number of	the fewest

Example: In 1992, Speedy Taxi Co. carried the fewest passengers among the four taxi services listed.

1. ______________________________

____

2. ______________________________

____

3. ______________________________

____

4. ______________________________

____

▶ Answers are on page 180.

READING DATA: USING ONLY THE INFORMATION GIVEN

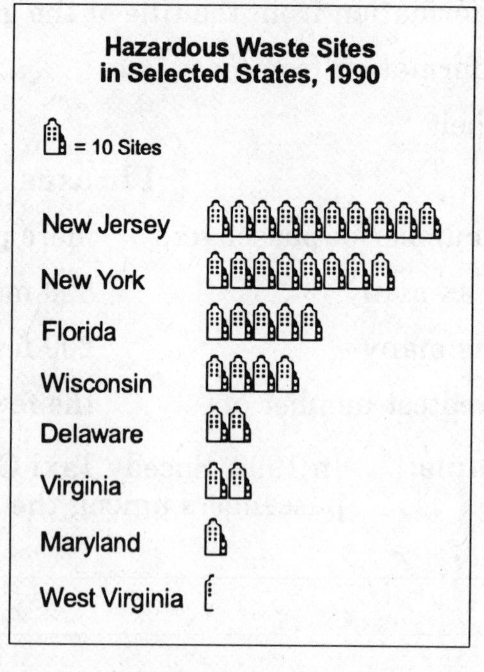

Source: Environmental Protection Agency

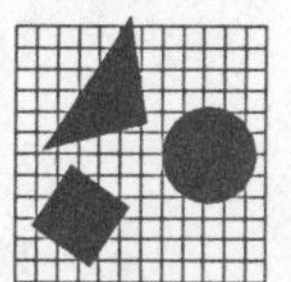

In the United States, what state had the greatest number of hazardous waste sites in 1990? Can you tell from this graph?

You might be tempted to answer "New Jersey" to the question above. Of the states *listed on this graph*, New Jersey does, in fact, have the greatest number. However, look closely at the title of the graph. Notice that only *selected* states are shown. A state that is not shown might have a greater number of hazardous waste sites than New Jersey. *The graph simply does not give us that information.*

In this lesson, you will work on using *only* the information on the graph.

■■ YOU TRY IT ▶▶

Make a statement about Wisconsin's hazardous waste sites using the data from the graph. Remember to include *only* information from

- the title
- the data given
- the key

Statement: ____________________

____

____

Does your statement sound something like this?

Wisconsin had 40 hazardous waste sites in 1990.

■■ YOU TRY IT ▶▶

The following statements about the graph *are not accurate*. Put a check mark next to the information listed below each statement that is *not* supported by the graph and therefore *cannot* support such a statement.

1. According to the Environmental Protection Agency, in 1990, Florida residents were more willing to store hazardous waste than the citizens of Virginia.

____ **a.** There were more hazardous waste sites in Florida than in Virginia in 1990.

____ **b.** Florida residents were willing to store hazardous waste.

____ **c.** Virginia residents fought hard against hazardous waste sites.

2. In 1990, Maryland, with 10 hazardous waste sites, was a more dangerous place to live than West Virginia with only 5 sites.

____ **a.** The sites in Maryland are better contained than those in West Virginia.

____ **b.** Maryland had more sites than West Virginia in 1990.

____ **c.** Maryland also has more crime than West Virginia.

3. West Virginia had the fewest hazardous waste sites in the United States in 1990.

____ **a.** West Virginia had 5 hazardous waste sites in 1990.

____ **b.** Hazardous waste sites are listed for all 50 states.

STATS

Did you know that . . . ? The Environmental Protection Agency keeps a *National Priorities List* that ranks the 56 most hazardous waste sites, *and 4 of the 10 most dangerous are located in New Jersey*.

You should have checked the following information that cannot be supported by the graph.

1. *b, c* We have no way of knowing how residents reacted from the given information.

2. *a, c* No information is provided about the condition of the sites or the crime rate for any of the states.

3. *b* Hazardous waste sites are listed for only 8 states.

▶ EXERCISE 3 Part One

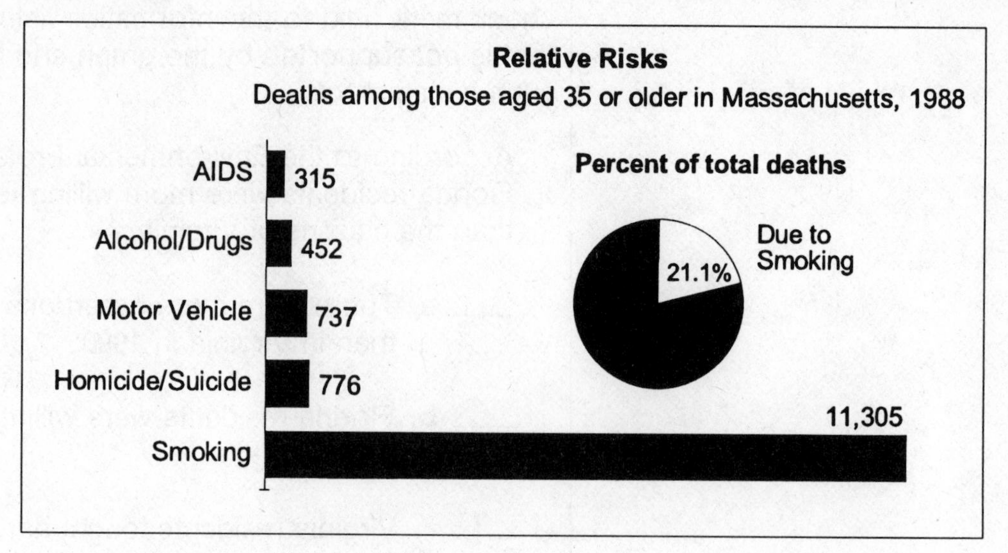

Source: Chronic Disease Surveillance Program, Bureau of Health Statistics, Research and Evaluation

State why each of the following is *not* an accurate statement about the data in the graph above.

Example: 315 people died from AIDS in 1988.

The graph refers only to Massachusetts (not to the world) and only to people 35 or older.

1. A total of 13,585 people aged 35 and over died in Massachusetts in 1988.

____

____

2. 11,305 more people would be alive in Massachusetts if cigarettes were illegal.

____

____

3. The abuse of alcohol and drugs is not as big a problem as smoking.

____

____

Part Two

Use the graph above to make two statements about the data. Be sure to use *only* the information given in the graph.

1. ______________________________

2. ______________________________

▶ Answers are on page 180.

THE LANGUAGE OF ESTIMATION

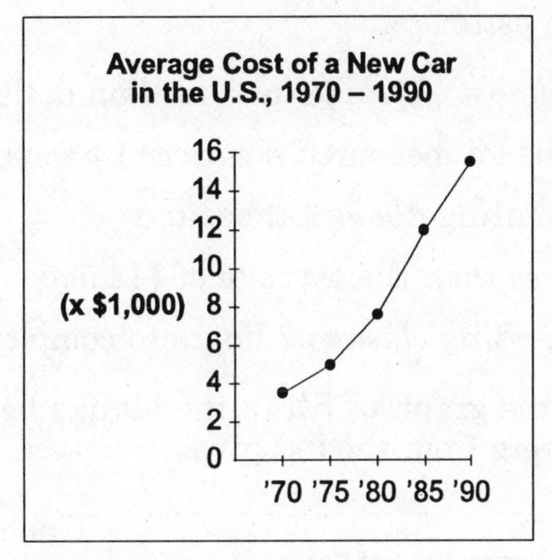

Source: Statistical Abstract of the U.S., 1991

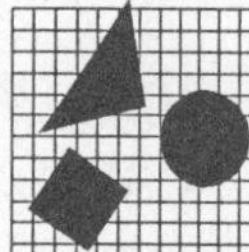

- What was the average car cost in the United States in 1990?
- Is it easy to determine from this graph? Why?

Many graphs are not meant to display exact figures. For example, on the graph above, it is impossible to tell the *exact* average car cost for any year. All you can really see is that the data point for 1990, for example, is somewhere between $14,000 and $16,000.

This doesn't mean that the graph is poorly constructed. Different graphs have different purposes. If you want to find a value on a graph like the one above, you'll need to **estimate**.

Estimate: to find a value *close to* the actual value

YOU TRY IT

There are many ways to express estimates. Use your knowledge of the words in dark type below to decide if each statement is true or false according to the graph above. Circle T or F.

T F **1.** The average cost of a new car in the United States in 1970 was **more than** $3,000.

T F **2.** The average cost of a new car in the United States in 1970 was **less than** $4,000.

T F **3.** The **approximate** cost of a new car in the United States in 1975 was $7,000.

1 and 2 You are correct if you found that the first two statements are true. The 1970 data point is higher than halfway between 2 and 4 (higher than 3) but below 4.

3 Did you find that this statement is false? The data point is between 4 and 6 (close to 5). Therefore, it is more accurate to say that approximate cost in 1975 was $5,000.

There are many ways in which estimates can be expressed. Look at the following list of words and phrases taken from a newspaper. All of them tell the reader that the number given is an estimate.

- *almost* a quarter of a million dollars
- the lot measured *just over* 13 acres
- totaling *about* a thousand
- *less than* the average of $14,500
- needing *close to* 7 hours to complete

▶ **EXERCISE 4** Use the graph to fill in the blanks below. Choose your answers from the list given.

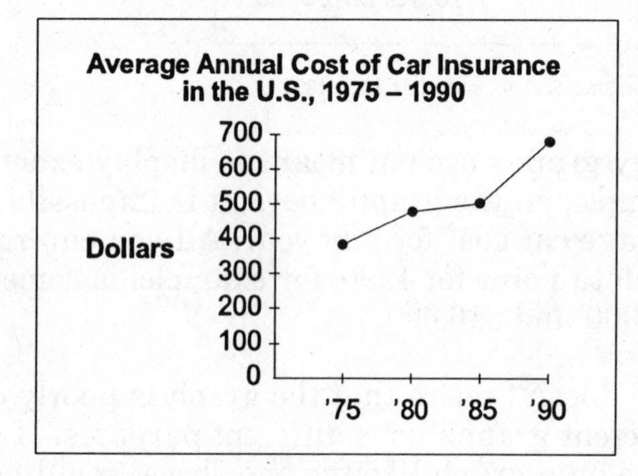

Source: Statistical Abstract of the U.S., 1991

Choose from this list:
less than
more than
almost
approximately
estimated to be
between

Note: More than one choice may be correct.

1. In 1980, the average annual cost of car insurance was ____________ $500.

2. The average annual cost of car insurance was ____________ $300 and $400 in 1975.

3. The difference in annual car insurance between 1985 and 1990 was ____________ $200.

Now make two statements using estimates of the data on the graph.

4. ____

____

5. ____

____

▶ Answers are on page 180.

ACCURATE ESTIMATING

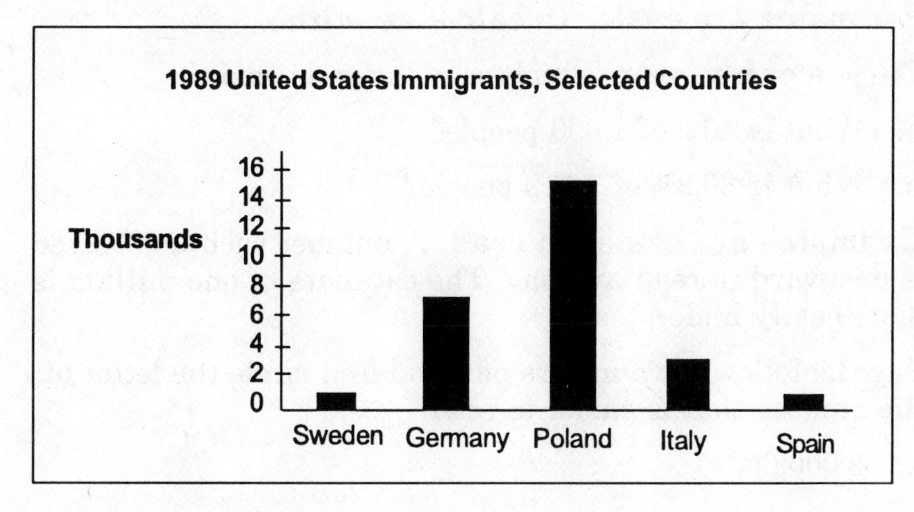

Source: Statistical Abstract of the U.S., 1991

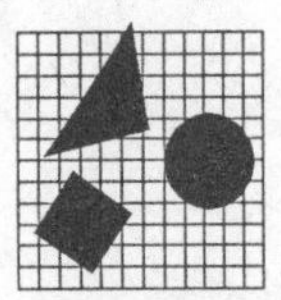

Read the two newspaper articles below. Do they contradict each other?

The Daily Journal

. . . in fact, the number of immigrants from Germany was **about 7,000** in 1989, **less than half** the number from Poland . . .

The Daily Express

. . . data indicates that **fewer than 7,000** German people immigrated to the United States in 1989, **about half** the number from Poland . . .

Which newspaper is reporting correct figures? Were there 7,000 immigrants from Germany or fewer? Did 15,000 Polish people immigrate to the United States, or 14,000?

Remember, *estimates are not exact*. The fact is that *both* newspaper articles above give acceptable estimates of the data. With estimates, 7,000 is considered close to half of 15,000.

■■ You Try It ▸▸

Estimate an answer to the problem at right. Do *not* find an exact number.

Write down your estimate before

3,675 + 199 = ____________
estimate

Which did you find as your estimate: 3,875, 3,900, or 4,000?

Whichever estimate you found, you are correct. In fact, any figure close to one of those above is a good estimate! Depending on how close to the actual answer you need to be, an estimate can have a wide range of values.

Why Estimate?

Why are estimates used so often? Here are some reasons:

1. **Estimates are easier to calculate with.**

   Circle **a** or **b** to show which calculation is easier.

   a. What is 40% of 1,000 people?

   b. What is 39.8% of 1,019 people?

2. **Estimates are easier to read.** A number such as 998,786 is awkward to read and say. The estimate of one million is more easily understood.

   Say the following numbers out loud and circle the letter of the number that is easier to read.

   a. 2,000,000

   b. 2,001,112

3. **Exact figures can change frequently.** The exact number of people living in Minneapolis, for example, changes every day. Estimates are used because they are as close as we can come to the actual number.

   Which of the following questions needs only an estimate for an answer?

   a. How many people are employed in the auto industry?

   b. How many passengers can the new Ford van carry?

4. **Exact figures are often not necessary.**

   As a consumer, which of the following would you want an exact figure for?

   a. the total cost of a microwave oven you want to buy

   b. the number of microwave ovens a store has in stock

Did you circle *a* for each question above? If so, you are understanding the purpose of estimation!

▶ EXERCISE 5 Part One

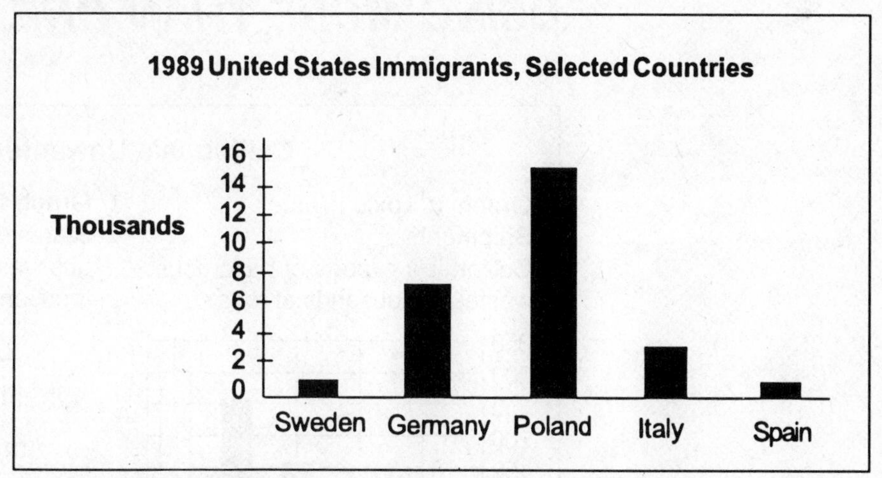

Source: Statistical Abstract of the U.S., 1991

Using the data, fill in the blanks that follow. Be sure to use words such as *about* or *approximately* when appropriate.

1. In 1989, immigration from Italy was __________ 3,000.
2. Immigration from Italy in 1989 was __________ the immigration from Spain.
3. The number of immigrants from Sweden in 1989 was approximately __________ the immigration from Spain.
4. In 1989, __________ more people immigrated from Poland than from Sweden.

Part Two

Look in a newspaper or magazine and find two places in which an estimate is used instead of an exact number.

1. How do you know an estimate is being used?

____

____

2. Why do you think an estimate is used instead of an exact number?

____

____

▶ Answers are on page 181.

Why Are Estimates Often Different?

Estimates can differ. Every time you estimate a number, you "have in mind" how close you want to be to the actual figure. Your estimate will depend on your purpose, or goal.

USING MORE THAN ONE DATA SOURCE

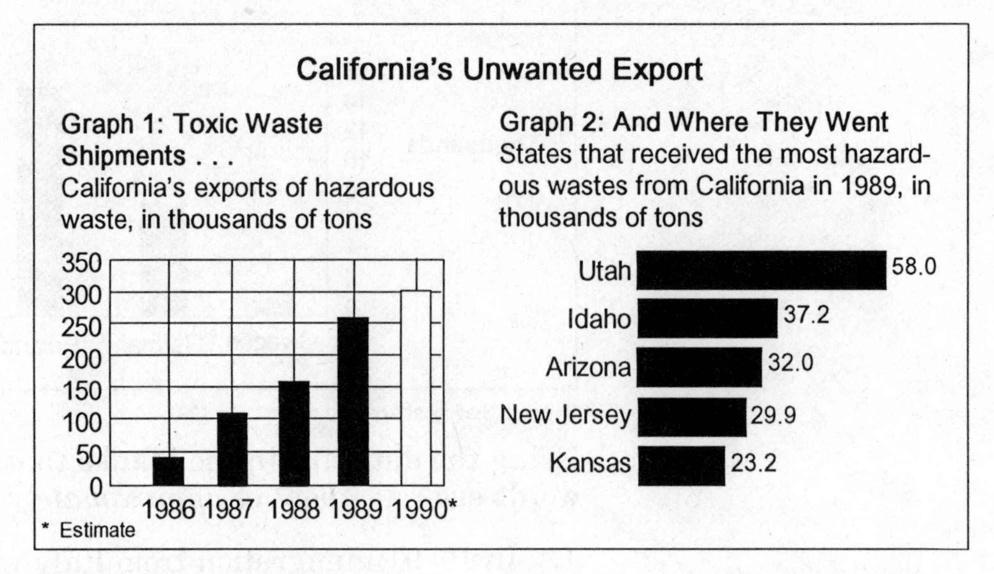

Source: California Department of Health Services

Source: California Health Department Toxics Program

- What is the purpose of the first graph above?
- What is the purpose of the second graph?
- What kinds of things can you learn from looking at both graphs together?

Complete the following statements using your data summary skills and filling in the what, where, when, and how information.

The first graph above shows ______________
what?

______________ ______________ ______________,
where? when? how?

according to the California Department of Health Services.

The second graph shows ______________
what?

______________ ______________ ______________,
where? when? how?

according to the California Health Department Toxics Program.

Your summaries should sound something like this:

The first graph shows **toxic waste shipments in California in 1986 through 1990 in thousands of tons**, according to the California Dept. of Health Services.

The second graph shows **the five states that received the most hazardous waste from California in 1989, in thousands of tons**, according to the California Health Dept. Toxics Program.

STATS

Did you know that . . . ? Japan generated 292 million tons of hazardous waste in 1989, and the United States generated 240 million tons. Compare these figures to Canada with 3 million tons and France with 2.

What information can you get by using *both* graphs that you can't get by using just one of them?

For example, the following question can be answered only if you use *both* graphs:

What *percent* of California's toxic waste exports was received by Utah in 1989?

The *first* graph tells you that California exported 250,000 tons of toxic waste in 1989. The *second* graph tells you that 58,000 tons went to Utah.

What percent of 250,000 is 58,000?

$58{,}000 \div 250{,}000 = .232 = 23.2\%$

Utah received 23.2% of California's toxic waste in 1989.

■■ YOU TRY IT ▶▶

Which of the graphs on page 88 do you need to answer these questions? Circle *Graph 1*, *Graph 2*, or *Both*. If the question cannot be answered by either of the graphs, circle *Neither*.

1. How many tons of hazardous waste did California export to Kansas in 1989? Graph 1 Graph 2 Both Neither

2. What fraction of California's toxic waste exports went to New Jersey in 1989? Graph 1 Graph 2 Both Neither

3. How many tons of toxic waste existed in Arizona in 1989? Graph 1 Graph 2 Both Neither

4. How many tons of toxic waste are estimated to be exported from California in 1990? Graph 1 Graph 2 Both Neither

▶ Answers are on page 181.

▶ EXERCISE 6

Milltown High School 1992 Age Breakdown

Total Student Population: 1,400

25% 15 years old
15% 17 years old
15% 13 years old
20% 14 years old
5% 18 years old
20% 16 years old

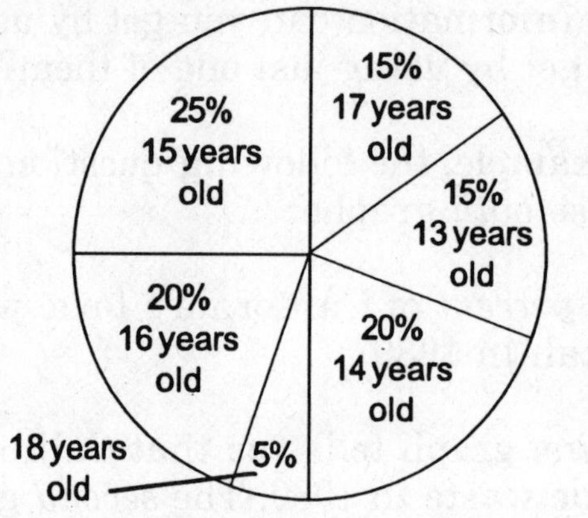

Students* Participating in School Activities

Activity	Number of Participants
Student government	210
Music/Band	140
Intramural sports	425
Library club	80
Social committee	120

*total student population: 1,400
Source: Milltown School Committee

For part *a* of each question, decide whether you need to use the graph, the chart, or both to answer the question. If the question cannot be answered using any of the information above, circle *Neither*.

For part *b*, answer the question if possible.

1. What fraction of the Milltown student population participated in music/band activities in 1992?

a. Graph Chart Both Neither

b. Answer: ____

2. Assuming that participation in all activities is not affected by age, how many 16-year-olds were in the Library Club?

a. Graph Chart Both Neither

b. Answer: ____

3. How many 13-year-olds attended Milltown High?

a. Graph Chart Both Neither

b. Answer: ____

4. How many girls played intramural sports at Milltown High?

a. Graph Chart Both Neither

b. Answer: ____

▶ Answers are on page 181.

PUTTING DATA IN DIFFERENT FORMS

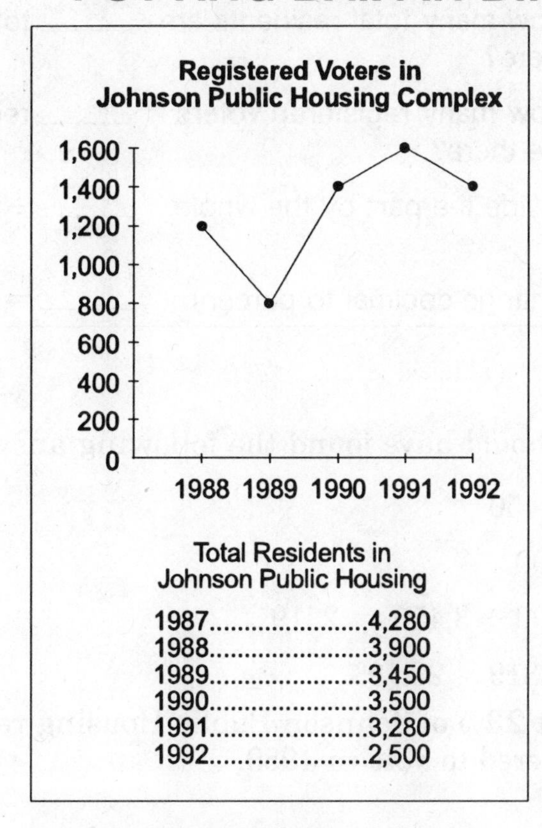

Total Residents in
Johnson Public Housing

Year	Residents
1987	4,280
1988	3,900
1989	3,450
1990	3,500
1991	3,280
1992	2,500

Source: League of Women Voters

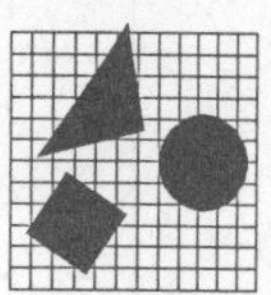

- What percent of Johnson Public Housing residents were registered to vote in 1990?
- Can you answer this question based on the information given?

From working with graphs and charts, you probably realized that the group of data above gives you *numbers*, not *percents*. Is there a way to change the form of the data so that you can answer the question?

Step 1. Find the total number of residents in 1990.

There were 3,500 residents. (the *whole*)

Step 2. Estimate how many people in the complex were registered to vote that year.

There were about 1,400 registered voters. (the *part*)

Step 3. Divide the *part* by the *whole*.

1,400 ÷ 3,500 = .40 or **40%** of Johnson Public Housing residents were registered voters in 1990.

● ● **MATH RECAP** ▶▶ To find out what percent one number is of another number, divide the *part* by the *whole*.

You Try It ▶▶

Use the data shown on page 91 to find what percent of Johnson Public Housing residents were registered to vote in **1989**.

1. How many total residents are there? ______ total residents (*whole*)
2. How many registered voters are there? ______ registered voters (*part*)
3. Divide the part by the whole. ______ (part) ÷ ______ (whole) = ______
4. Change decimal to percent. ______ = ______%

You should have found the following answers:

1. 3,450
2. 800
3. $800 \div 3{,}450 = .2319$
4. $.2319 = 23.19\%$

About **23%** of Johnson Public Housing residents were registered to vote in 1989.

Why Change the Form of Data?

Why is it useful to see the data on page 91 changed to percent? The following activity will help you see why.

1. Compare the **number** of registered voters in 1992 to those in 1991.

   Did the number go up or down?

   The number of registered voters in the Johnson Public Housing Complex **fell** from 1,600 in 1991 to 1,400 in 1992.

2. Why do you think this drop occurred? Put a check mark next to any reasons that seem sensible to you.

   ☐ fewer volunteers working to register people

   ☐ no appealing candidates for office in 1992

   ☐ apathy or resentment among residents

   ☐ other: ______________________

3. On the graph, plot the *percents* of registered voters as determined from the data on page 91 and in the box below.

Remember: to find percent, divide the part (number of registered voters) by the whole (number of residents). Percents for 1988 and 1989 have been computed for you.

STATS

Did you know that . . . ? The highest percent of registered voters who actually voted in the 1988 presidential election was 80.9% in Connecticut. The lowest percent was in Oklahoma, at 53.3%.

1988: 1,200 ÷ 3,900 is about .31, or **31%** (Estimate this as 30% on the graph.)

1989: 800 ÷ 3,450 is about .23, or **23%**

1990: 1,400 ÷ 3,500 is .40, or ______%

1991: 1,600 ÷ 3,280 is about .49, or ______%

1992: 1,400 ÷ 2,500 is .56, or ______%

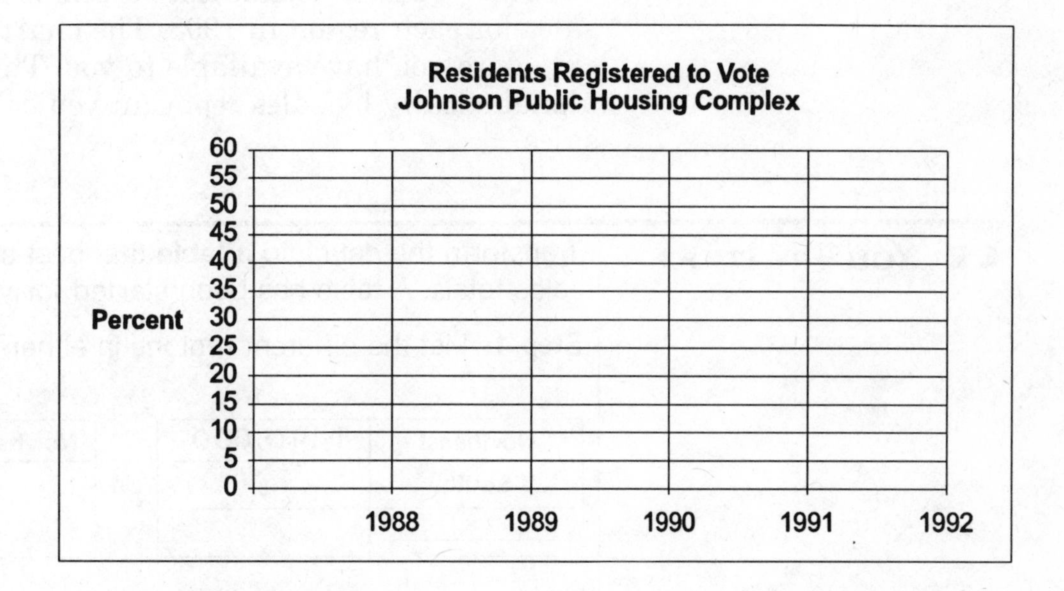

4. a. What happened to the percent of registered voters between 1991 and 1992? Did the percent rise or fall?

b. Why do you think this happened? ______________________

____

▶ Answers are on page 181.

Perhaps you already figured out that the number of registered voters fell because the *total number of residents* fell as well.

Changing the form of data can help you see things that might not be obvious at first.

Here's another example of putting data into different forms.

Territory Sales Figures, 1993

Sales Representative	Region	Sales in Dollars
David Burns	Northeast	$475,000
John Fay	South	290,000
Lisa Gagnon	Northeast	325,000
Kirsten Hutchinson	Northeast	190,000
Tamara Jenkins	South	315,000
Julio Kim	Midwest	219,000
Juanita Nalda	Midwest	220,000
Laura Rice	South	110,000
Rena Sanchez	South	328,000
Herm Thomas	Midwest	200,000
Raul Whitney	Northeast	150,000

Suppose you need to calculate a company's total dollars in sales for each region in 1993. The chart above contains the only data you have available to you. The chart is organized alphabetically by sales rep. Can you calculate total regional sales?

■■ YOU TRY IT ▶▶

Transform the data into a **table** that best summarizes the regional sales totals. A table has been started for you.

Step 1. List the different regions in either a column or a row.

Northeast	1,140,000
South	
______	

OR

Northeast	______	Midwest

Step 2. Add up the sales figures for each region.

Northeast: 475 + 325 + 190 + 150 (× 1,000) = **1,140,000**

South: 290 + 315 + 110 + 328 (× 1,000) = ______________

Midwest: 219 + 220 + 200 (× 1,000) = ______________

Step 3. Finish writing the totals in the appropriate column or row.

▶ Answers are on page 181.

Although the original purpose of a set of data might be different from your purpose, changing the *form* of the data can often help you get the information you need.

▶ **EXERCISE 7** Change the form of the data by following the steps indicated.

1. Draw a chart that shows what **percent** of patients listed below have each different insurance plan. Your information source is this graph.

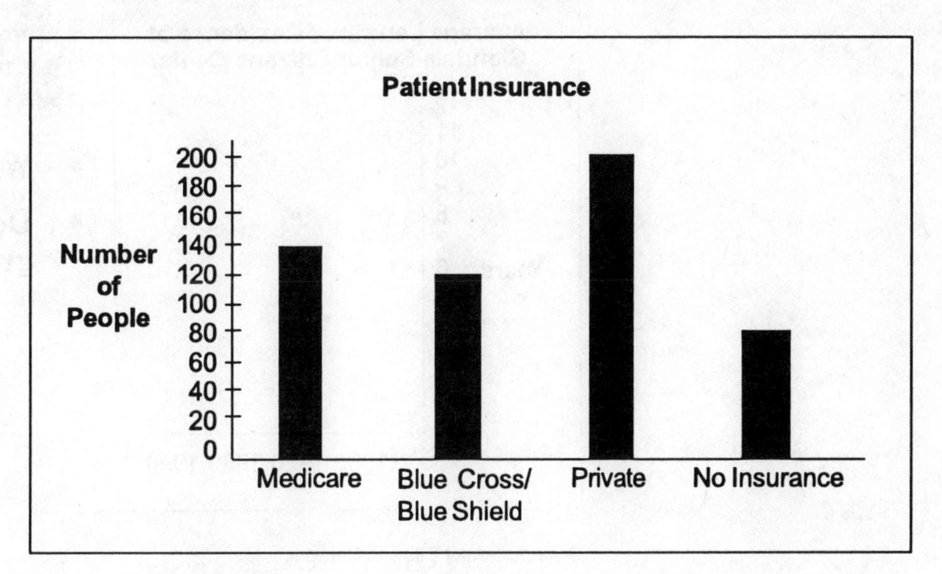

Step 1. Find the total number of patients by adding up the numbers from each insurance category.

Total Patients: ____

Step 2. To find each percent, divide the part (each insurance category) by the whole (total patients).

Step 3. Fill in the chart started below.

Insurance	Medicare	Blue Cross/ Blue Shield	Private	No Insurance
Percent of Patients				

2. An agency estimates that the total number of homeless people in Sun County is 9,500. Use this and the following information to answer the questions.

Sun County Homeless Population

Area	Percent
Downtown Area	37%
Central Square	29%
East Side	22%
West Side	10%
Outlying	2%

a. Figure out the *number* of homeless people that were counted in each area.

b. Make a chart that presents this information.

▶ Answers are on page 181.

SEEING TRENDS/MAKING PREDICTIONS

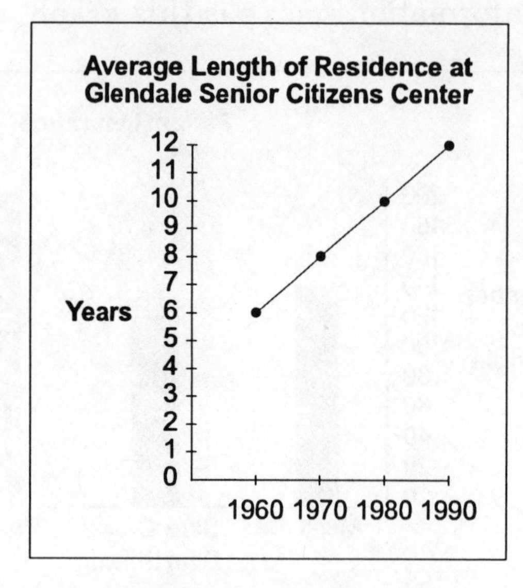

Source: Senior Citizens Network

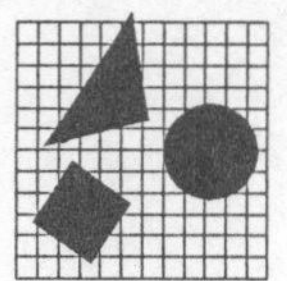

- What is a trend?
- Do you see a trend on the graph at left?

Trend: a general movement following a certain pattern or course

To learn what a **trend** is, fill in the blanks:

- Between 1960 and 1970, the average length of residence at Glendale Senior Citizens Center rose by ___ years.
- Between 1970 and 1980, the average length rose by ___ years.
- Between 1980 and 1990, the average length rose by ___ years.

Did you notice something repeating? Every ten years, the average length of residence goes up by two years. The graph above indicates a pattern, or **trend**, in average length of residence at Glendale Senior Citizens Center.

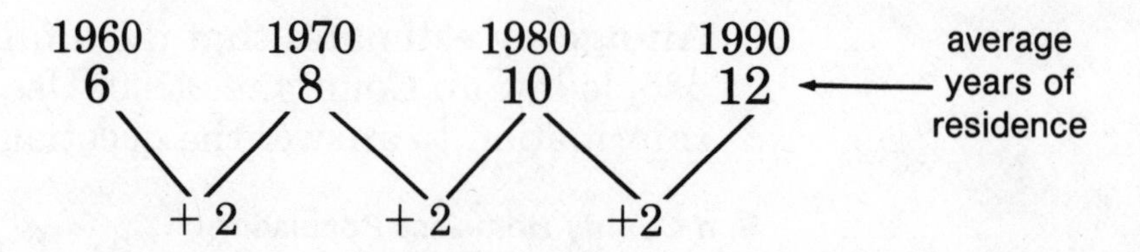

If the trend continues, what will the average length of residence be in the year 2000?

To figure this out, follow the trend by adding two years to the average length of residence in 1990:

CRITICAL THINKING

In 1990, the population of Americans over the age of 65 was almost 30 million, a 24.1% increase over the 1980 figure.

Why do you think this is true?

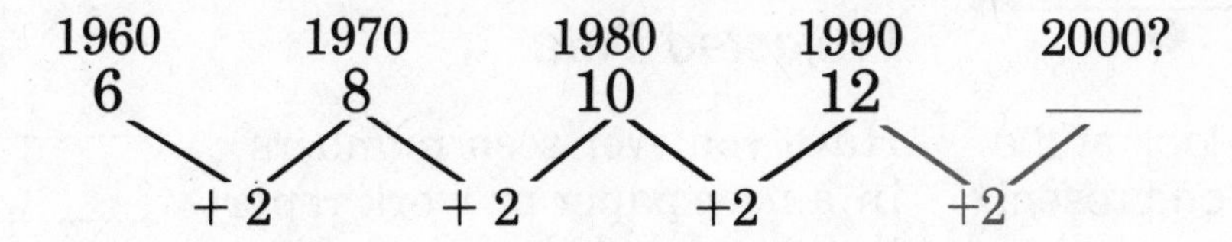

The average length of residence in the year 2000 will be **14** (12 + 2), *if the trend continues*.

General Trends

The trend on the graph on page 96 showed a rise of exactly the same number of years every ten years (or every decade). Sometimes trends are more general—they follow a pattern, but one that is not as simple as the trend you saw before. Let's look at an example.

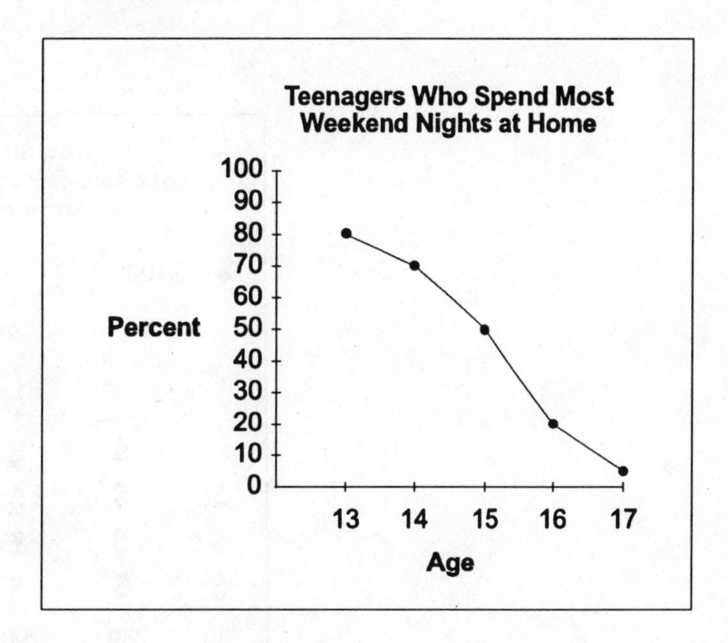

Do you see a trend concerning teenagers who stay home most weekend nights?

As you can see, there is certainly a *downward* trend with increasing age, but the size of the drop is different at each jump in age.

If the trend continues, will the percent of 18-year-olds who stay at home most weekend nights be *higher* or *lower* than the percent of 17-year-olds?

You're right if you said **lower**. Although you cannot tell how much lower, you can see from the **general trend** of the data that the percent will go down.

Projected Data

Projection: a look at the future based on present trends or data

Have you ever seen numbers in a newspaper or work report that are labeled *projected*? This label means that the data is not yet available. Instead, experts are making an estimate based on a trend. The chart shows **projections** of the world's population in the next three decades. These estimates are based on population totals measured in previous decades.

World Population*	
Year	**Population**
2000	6,291,000,000
2010	7,255,000,000
2020	8,281,000,000

*Projected
Source: U.S. Bureau of the Census

▶ **EXERCISE 8** Use the data below to solve problems 1–2.

Approximate Monthly Rent for a Two-Bedroom/Two-Bathroom Apartment in Del County

$ = $100

'88 '89 '90 '91 '92 '93

1. If the trend shown continues, approximately how much would the monthly rent for a two-bedroom/two-bathroom apartment be in 1994 in Del County?

   $ ________

2. Make a statement about the approximate monthly rent in 1995.

   ____

   ____

Use the list of phrases below to write four statements about the data displayed on the graph below.

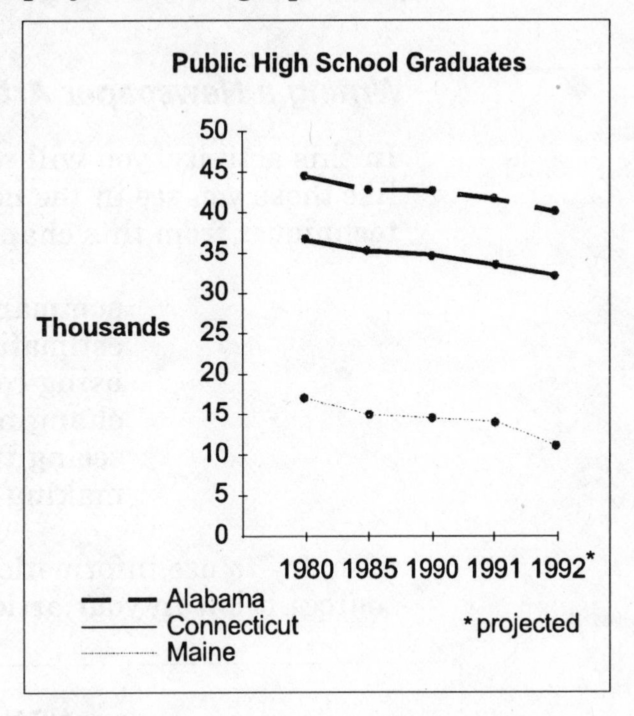

Source: U.S. National Center for Education Statistics

if the trend continues	was less in 1990
dropped steadily	approximately 20,000 more
about 30,000 fewer	is projected to be

Example: The number of public high school graduates in Connecticut was less in 1990 than in 1985.

3. ______________________________

4. ______________________________

5. ______________________________

6. ______________________________

STATS

Did you know that . . . ?
Every 2 seconds, 9 babies are born and 3 people die. The net increase of 3 people each second results in a growth in world population of
- 10,800 per hour
- 259,200 per day
- 1.8 million per week
- 7.7 million per month
- 93 million per year

▶ Answers are on page 182.

Writing a Newspaper Article

In this activity, you will write articles and letters to the editor like those you see in the newspaper. Try to use as many of the techniques from this chapter as you can, including

summarizing
estimating
using correct information
changing the form of data
seeing trends
making predictions

Also, try to use information from two or more of the data sources below in your article.

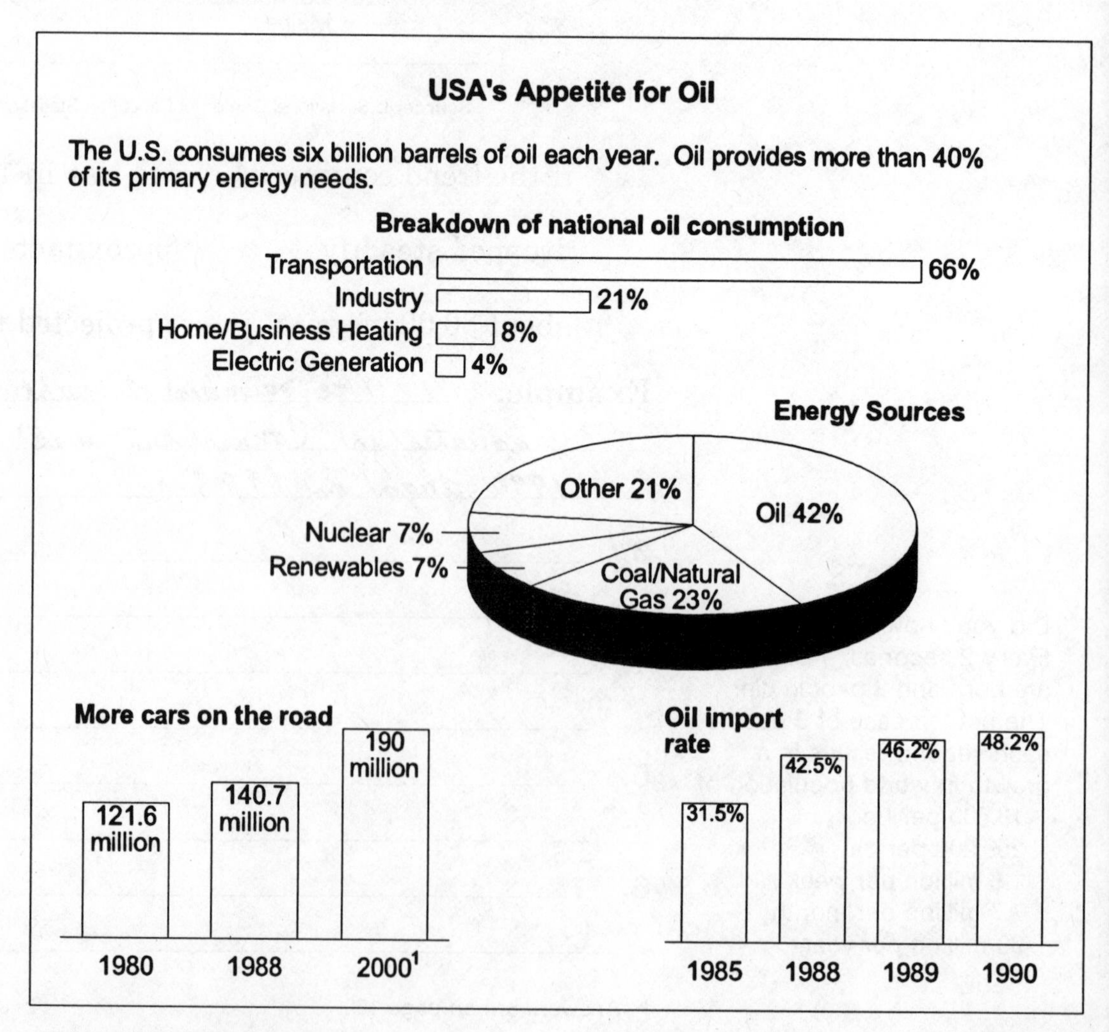

Source: U.S. Department of the Interior

[1] projection

Writing Activities

1. Use one of the following statements as the first sentence in an article about our national energy situation. Write the article using any of the data on page 100.

   Choose one:

   a. The United States uses most of its six billion barrels of oil on transportation each year.

   b. The projected number of cars on the road in the United States in the year 2000 is about 35% higher than in 1988.

   c. Oil represents almost half of all of our energy sources.

2. Do you have an opinion about our country's energy usage? For example, some people think that we are using up too much of our natural resources and that we should use more nuclear energy or renewable energy sources such as solar or wind power. Other people believe that nuclear power is too dangerous and should not be used at all. Still other people believe that, no matter what our energy source is, we use too much of it on transportation (such as cars).

   On a separate sheet of paper, write a letter to the editor of a newspaper stating your opinion on energy in the United States. Use statistics to back up your point of view.

GROUP PROJECT

Summarizing Your Group Project Data

Choose some of the data you collected and organized in Chapter One and Chapter Two and plan to summarize what you have discovered.

Write an article that answers *one* of the following questions, using your data as support for your ideas. Or, you may answer a question of your own if you prefer. In your article, be sure to

- summarize your data accurately
- use specific numbers or percents
- try showing your data in different forms
- notice trends
- use any other data or information you've found in books or newspapers

Questions

1. Do people who live alone tend to watch more television than those who live with others?
2. Are married people more often "day people" or "night owls"?
3. Do more people recycle glass or paper? How about aluminum?
4. Are people today satisfied with their president?
5. Do people with fewer close friends go out more or less often than people with more close friends? Which group watches more television?
6. Is the economy or the environment a bigger concern for people?
7. Who tends to have pets—married people or single people? male or female?

8. What section of the newspaper is the most popular? Do more men than women read the sports first?

9. Who thinks Americans should buy only American products—people who are concerned about the economy or people who are concerned about the environment?

10. Other: ______________________________

____

CHAPTER FOUR

Using Probability and Statistics

WHAT ARE THE CHANCES?

"I'll probably get put on the late shift at work," said Dan, a new employee. "Personnel told me that 9 out of 10 new people start out there."

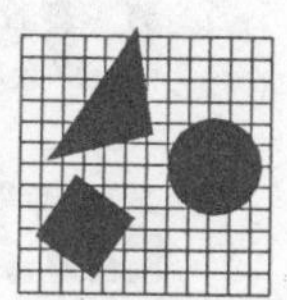

- Why does Dan think he'll be put on the late shift?
- Is Dan *sure* what shift he'll be put on?

Probability: the chances that an event will occur

To picture **probability**, use coin tossing. Toss a coin in the air and let it fall. Is one side more likely to be faceup?

YOU TRY IT

1. Toss a coin ten times and record the number of heads and tails.

   First ten tosses:
   Heads ________
   Tails ________

2. Now toss the coin ten more times and again record the results.

   Second ten tosses:
   Heads ________
   Tails ________

3. Write a statement about the total number of heads vs. the total number of tails.

   ____
   ____

You can show the results of the coin-tossing experiment as a fraction or a ratio:

$$\frac{\text{number of times outcome occurred}}{\text{total number of outcomes}}$$

Using the information from above, write the fraction showing the number of *tails* in *20* tosses.

$$\frac{\square}{\square} \quad \begin{array}{l}\text{number of tails}\\ \text{total tosses}\end{array}$$

Fractions or ratios such as these can be used to *estimate the probability* of tossing either heads or tails. In fact, the more times you toss the coin and record the outcome, the closer you will come to knowing the actual probability. By now, you may realize that the chance of tossing heads (or tails) is close to $\frac{1}{2}$.

Here's how we represent probability:

$$\text{Probability of an event} = \frac{\text{number of desired outcomes}}{\text{total number of possible outcomes}}$$

In the case of the probability of tossing heads, the "desired outcome" is heads. Therefore, the probability of tossing heads is expressed as:

probability → $P(H) = \frac{1}{2}$ ← 1 (because there is 1 head on the coin); 2 (because there are 2 possible outcomes, heads or tails)

↑ heads

Go back and compare this fraction to the other fractions you have worked with in this experiment. What do you notice?

When you are dealing with coin tossing, all fractions and probabilities are very close to $\frac{1}{2}$.

▶ EXERCISE 1

You have learned that probability depends on the number of *desired outcomes* and the total number of *possible outcomes*. Apply what you know to this situation. **Write each probability as a fraction.**

When you roll a die, the chances are the same that any one of the numbers 1 to 6 will appear faceup.

Example: What is the probability that a six will be faceup?

$\frac{1}{6}$ (there is one six on the die) / (number of sides on the die)

1. What are the chances of rolling a one? $\frac{\square}{\square}$ (number of ones) / (number of sides)

2. What is the probability of rolling an odd number? ______

3. What are the chances of rolling a number divisible by 3? ______

Now let's return to Dan's situation from the beginning of this lesson.

4. Write a fraction representing the probability that Dan will be put on the late shift. ______

5. What is the probability that Dan will *not* be put on the late shift? ______

▶ Answers are on page 182.

PROBABILITIES BETWEEN 1 AND 0

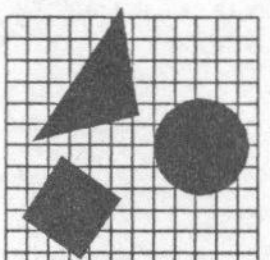

- What is the probability that you will fly to the moon tomorrow?
- Express this probability as a fraction.

$\frac{0}{1}$ ← number of outcomes / number of possible outcomes
(There's only one possibility: you won't go.)

The chances that I will fly to the moon tomorrow are 0.

What is the probability that tomorrow will follow today?

$\frac{1}{1}$ ← number of outcomes / number of possible outcomes
(Tomorrow always follows today.)

It is virtually a "sure thing" that tomorrow will follow today. **Therefore, we can express this probability as 1.**

- A probability of 1: the event will certainly happen.
- A probability of 0: it's not possible for the event to occur.
- All other probabilities are expressed as fractions between 0 and 1, just as you learned in the last lesson.

YOU TRY IT

Imagine that the numbered cards below are *facedown* on a table. You pick up one card, not knowing which you are choosing.

2	4	6	8

1. What is the probability that the number on the card you chose is a 4? ______

2. What is the probability that the number on the card is greater than 3? ______

3. What is the probability that the number on the card is divisible by 2? ______

4. What is the probability that the number on the card is greater than 8? ______

▶ Answers are on page 182.

▶ EXERCISE 2

Part One

Look at each of the following situations and circle whether it has a probability of 0 (no chance of occurring), 1 (a sure thing), or somewhere in between.

What are the chances that . . .

1. you will eat a hamburger in the next year?

   0 in between 1

2. it will rain somewhere in the world today?

   0 in between 1

3. someone in your state will die tomorrow?

   0 in between 1

4. you will win a million dollars in the lottery?

   0 in between 1

5. a plane is taking off now?

   0 in between 1

6. we will have world peace tomorrow?

   0 in between 1

7. a woman is giving birth right this minute?

   0 in between 1

8. the sun is shining somewhere in the world?

   0 in between 1

9. everyone will use public transporation by the year 2000?

   0 in between 1

10. the Cubs will win the World Series?

    0 in between 1

STATS

Did you know that . . . ?

The probability that an American will eat breakfast in the car is $\frac{1}{4}$.

The probability that an American will skip breakfast is $\frac{1}{10}$.

Source: *Harper's* magazine

Part Two

1. Write down three events that have a probability of 0. Use your imagination!
2. Write down three events that have a probability of 1. Be creative.

▶ Answers are on page 182.

USING DATA TO ESTIMATE PROBABILITY

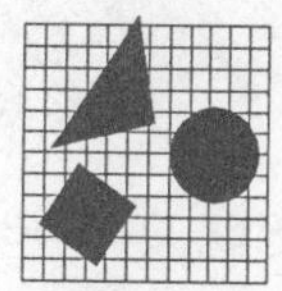

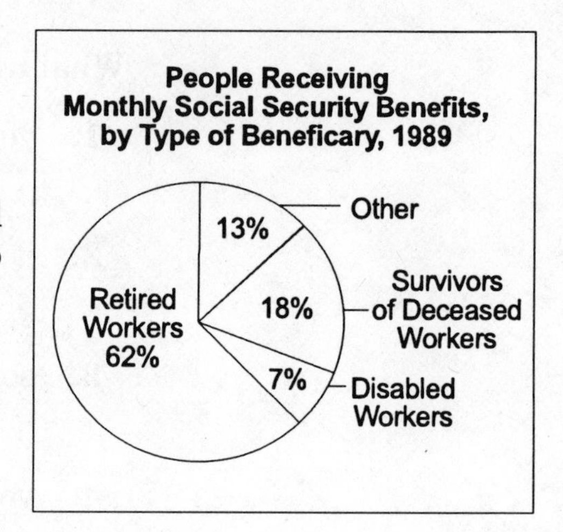

Suppose you work in a Social Security office. A person who receives Social Security benefits approaches your desk. What are the chances that this person is a retired worker?

Look at the data above. The chances that a person who receives Social Security benefits is retired are pretty good—a little more than $\frac{3}{5}$ or 3 in 5. How do you know this?

62% of people receiving benefits are retired.

$\frac{62}{100}$ retired people / all people receiving SS benefits

● ● MATH RECAP ▶▶ Remember that 62% means 62 out of 100.

This relationship can also be expressed in terms of probability:

Out of 100 people receiving SS benefits, 62 are retired.

$\frac{62}{100}$ is approximately equal to $\frac{60}{100}$ is equal to $\frac{30}{50}$ or $\frac{3}{5}$

Out of every 5 people receiving SS benefits, approximately 3 are retired.

Therefore, you can say that the chances are about 3 in 5 that the person who approached your desk is a retired beneficiary.

■■ YOU TRY IT ▶▶

What are the chances that the person who approached your desk is *not* a retired beneficiary?

1. According to the graph, what percent of people receiving benefits are *not* retired?

   100% − 62% = ______

2. Express this percent as a fraction, estimate, and reduce.

   $\frac{\square}{100} \approx \frac{\quad}{\text{estimate}} = \frac{\quad}{\text{reduced}}$

3. Express the fraction in probability terms.

   The chances that a person receiving SS benefits is *not* a retired beneficiary are about

   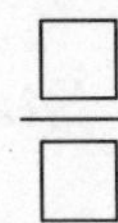

Note: Since you already know that the chances are 3 in 5 that a person *is* a retired beneficiary, you could also subtract to find that the chances are 2 (5 − 3) in 5 that a person is *not* a retired beneficiary.

Does your statement sound something like this?

The chances that a person receiving SS benefits is not a retired beneficiary are **about 2 in 5**.

How Can We Be Sure?

Can you correctly identify what type of beneficiary is approaching your desk simply by using probability?

No, you can't. Even if the chances are 9 in 10 that a person is retired, you could never be sure that the person approaching your desk is not that 1 person in 10 who *isn't* retired.

As you learned in the last two lessons, probability allows you to make *reasonable guesses*, not exact predictions.

For example, it is very hard to predict the outcome of one coin toss. (Will it come up heads or tails?) However, probability allows you to reasonably guess how many heads (or tails) would come up in 100 tosses.

Similarly, the data above cannot assure you that the person approaching your desk is a retired beneficiary. However, if 100 people receiving SS benefits approached your desk, you could make a reasonable guess that somewhere around 62 of them would be retired beneficiaries.

▶ EXERCISE 3

Use the data to determine the probabilities below.

Registered Voters, Kensington County		
	1991	1992
Democrats	1,280	1,580
Republicans	960	820
Independents	320	600

Example:
What were the chances that in 1992 a registered voter in Kensington County was an independent?

Step 1. How many 1992 registered voters were there in all?

$$1{,}580 + 820 + 600 = 3{,}000$$

Step 2. What fraction was registered independent?

$$\frac{600}{3{,}000} = \frac{1}{5}$$

The chances that a voter registered as an independent in Kensington County in 1992 are **1 in 5**.

1. What were the chances in 1991 that a voter registered in Kensington County was a Democrat?

Step 1. How many 1991 voters were there in all?

____ + ____ + ____ = ____

Step 2. Write a fraction comparing the number of 1991 registered Democrats to total 1991 registered voters in Kensington County.

2. What was the probability that a voter registered in Kensington County in 1991 was *not* a Republican?

____

3. What was the probability that a voter registered in Kensington County in 1991 was an independent?

____

4. Was the probability that a voter registered in Kensington County in 1992 was *not* a Democrat less than $\frac{1}{2}$ or greater than $\frac{1}{2}$?

____

5. Were the chances that a voter in Kensington County in 1992 was a registered Republican less than $\frac{1}{3}$ or greater than $\frac{1}{3}$?

____

▶ Answers are on page 182.

AN INTRODUCTION TO SAMPLING

Advertising Executive: "We took a survey of 1,500 people, and 1,125 of them said that they did not like our chocolate french-fry idea. So we've decided not to sell them at all."

French-Fry-Loving Friend: "Just because 1,125 people out of this entire country don't like the idea doesn't mean the rest of us don't! Why don't you listen to us instead?"

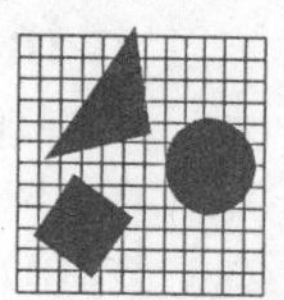

- How do you think a company decides whether or not a product will sell well?
- What mistake is the friend making in his comment at left?

To the french-fry lover, 1,125 people does not seem like a large number—especially since he compares the number to the population of the whole country! However, the company is actually comparing **1,125** to the **1,500** people it surveyed—not all 250 million people in the United States.

YOU TRY IT

Use what you have learned about percents and probability to complete the following statements:

The chances that a person would not like chocolate-covered french fries are 1,125 out of _____, or 3 in _____.

The percent of Americans who do not like the idea of chocolate-covered french fries is _____%.

In fact, based on the company's survey, the chance is **3 in 4** that a person would not like chocolate-covered french fries—a total of **75%**. With so many people disliking the idea, the company was very wise not to go ahead and sell them.

But how can the company base its decision on the responses of 1,500 people? The company can use a **sample** of the population to predict how a larger group will respond.

Population and Sample

The study of statistics makes it possible to learn about a large group of people, places, or things **(the population)** by looking at information about a smaller part of that group **(the sample)**. Look at some examples of how samples are used to make decisions and predictions every day in government, the business world, and even your own life:

Population: the larger group you want information about

Sample: a smaller part of the *population* that you actually get information *from*

Everyday Data
When choosing where to shop, people often compare prices in different stores. For example, suppose one store charges more for aspirin, paper products, toothpaste, and diapers. "This store has higher prices," you might say to yourself. "I'm not going to shop here." Based on a *sample* of prices, you made a decision about the prices in general, the *population*.

Business
A factory worker inspects a carton of light bulbs that has just been packed. She finds that 18 of the 240 bulbs are broken. "Something is wrong with the packing machine on this line," she says. "We'll need to recall all cartons that were packed today." Based on a *sample* of bulbs, the worker makes a prediction about what she'll find in all the cartons, *the population*.

Government
Government statistics indicate that population in the Northeast grew to fifty-one million in 1990. Did the government count each and every person living in the Northeast? Absolutely not. Instead, it chose a *sample* of the region, and based on how the population grew in that sample, it made an accurate estimate of the entire Northeast *population*.

▶ **EXERCISE 4** For each situation given below, write down the *population* and the *sample*.

Example: To predict the total number of votes cast for each presidential candidate, a television station interviews voters and collects a sample of their preferences for president.

Population: total votes ______ Sample: voters interviewed

1. An education official wants to know the standardized test scores for sixth-graders in his state. He looks at records from sixth-graders in six different towns in the state.

   Population: ______________ Sample: ______________

2. A candidate for mayor would like to find out how she is perceived among low-income residents in her city. She asks her staff to survey people living in city low-income housing units.

   Population: ______________ Sample: ______________

3. A shift supervisor counts defective aluminum canisters produced between 12:00 and 1:00 A.M. to discover how many defective canisters were produced on the entire early-morning shift.

   Population: ______________ Sample: ______________

4. A customer returns three poorly made toys to the same toy store and decides that none of the store's merchandise is of good quality.

   Population: ______________ Sample: ______________

5. A store that sells cookies offers free pieces of its chocolate chunk cookies to passersby. People who taste the cookies decide that all the baked goods in the store are delicious.

   Population: ______________ Sample: ______________

▶ Answers are on page 182.

RANDOM SAMPLES

Optimistic Moviemaker: "I want to make a movie that will appeal to *everyone*—a real blockbuster. So, my partners and I interviewed more than 100 people, and the results were amazing. Over 95% of the people we interviewed wanted to see more horror movies—more violence, more monsters, more creepy stuff. It's obvious what kind of movie we should make!"

Money-Maker: "You didn't by any chance interview the 100 people leaving the 4 P.M. showing of that new horror picture downtown, did you?"

Optimistic Moviemaker: "Why, yes, we did. How did you know?"

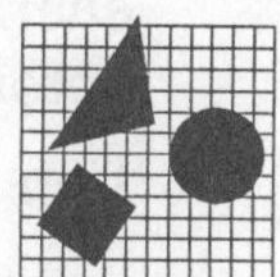

- What is the *population* being discussed at left?
- What is the *sample?*
- Do you think most Americans want to see more horror movies?

Representative: having the same general characteristics as the population

To make accurate decisions and predictions based on statistics, you must use a sample that is **representative** of the population you want to learn about. Is this true of the sample taken above?

YOU TRY IT

Put a check mark next to the categories of people *you* think would be likely to watch a weekday afternoon horror movie downtown.

- ☐ Teenagers
- ☐ Adults
- ☐ People who work during the day
- ☐ People who do not work during the day
- ☐ City dwellers
- ☐ Suburbanites
- ☐ People who like horror movies
- ☐ People who don't like horror movies

Did you find that the people our optimistic moviemaker interviewed were probably young people who were not working during the day and who generally like horror movies? The sample might include exceptions, but these would probably be the characteristics of most of the people interviewed.

What does this tell you about the sample chosen for the moviegoing population? Certainly, it was not representative of the public in general.

A representative group of Americans should include *at least*:

both males and females
people of different ages
people of different races and ethnic groups
employed and unemployed people
people from cities, suburbs, and rural areas
poor, middle-class, and wealthy people

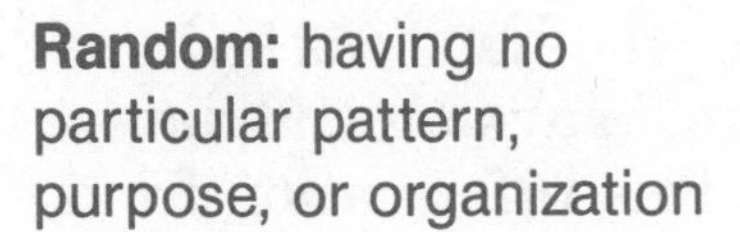

How Can We Get a Representative Sample?

Random: having no particular pattern, purpose, or organization

There are different ways to obtain a representative sample of a population. The best-known and generally most accurate method is called **random sampling**.

Suppose you had a jar full of colored balls—some red, some blue, and some green—all exactly the same size. If you mixed them all up in the jar, closed your eyes, and pulled out a number of them, you would have a *random sample* of the balls in the jar. None of the balls is more likely to be pulled out than another.

STATS

Did you know that . . . ? An accurately obtained random sample of 1,500 Americans can provide information about the entire United States population of 250 million people.

A sample of a population is considered a random sample if each member of the population is *equally likely to be chosen*.

When a government, business, or organization wants information about a population, it uses random sampling to gather data. A population can be any group—not just humans. It can include

- all mammals living in Southeast Asia
- children under age five with the same flu virus
- graham crackers produced daily in New York State

Remember that **population** refers to *the group that we want information about*. To get this information, we look at a *random sample* within this population.

▶ **EXERCISE 5** First find the population and the sample in each of the following situations. Then decide why each sample is *not* representative of the desired population.

1. A government official wants to find out what Americans think about cuts in the U.S. defense budget. He interviews 200 people outside a shopping mall in a small town near an army base.

Population: ______________ Sample: ______________

What is wrong with the sample? ______________

____

2. A consumer wants to shop at the market with the freshest food. She reads the "date packed" label on whole chickens in five different stores.

Population: ______________ Sample: ______________

What is wrong with the sample? ______________

____

3. "The people of the United States want a president who does not put a tax burden on people who have worked hard to achieve the American dream," says a candidate for office. The candidate's staff got this information from a survey of members of the American Small Business Owners Association.

Population: ______________ Sample: ______________

What is wrong with the sample? ______________

____

4. A production worker inspects the first 100 circuit boards that are manufactured one Tuesday morning. She wants to know how many defective boards the manufacturer produces each month.

Population: ______________ Sample: ______________

What is wrong with the sample? ______________

____

▶ Answers are on page 183.

AN INTRODUCTION TO SURVEYS

- A candidate for public office wants to know whether or not the people in his state favor the death penalty.
- The personnel director at a large corporation wants to know if employees are happy in their jobs.
- A market research specialist wants to find out whether a company could make money manufacturing disposable socks.

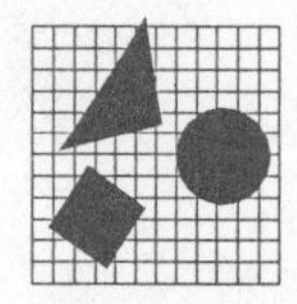

What can these people do to get the information they need?

Survey or **Poll:** a systematic collection of data from a random sample of a population

Government agencies, businesses, schools, political candidates, and nonprofit organizations are just some of the individuals and groups that conduct **surveys** or **polls** to obtain information.

■■ YOU TRY IT ▶▶

Have you ever been involved in or heard about any surveys? What were they about? Have you ever phoned in your response to a televised survey question?

___

___

___

Surveys are usually conducted in one of three ways. Let's take a look at how each works. Put a check mark next to the survey you would be most likely to fill out and return for each survey category.

1. Mail

Surveys by mail require you to fill out a questionnaire and mail it back to the sender.

☐ Magazine music survey (favorite group, favorite song, etc.)

☐ Warranty and questionnaire that come with a product (stereo, appliance, etc.)

☐ Survey that comes with a cereal sample packet

Would a self-addressed, stamped envelope or card influence your decision to respond to a survey? How?

2. **Telephone**

Currently, there are two types of telephone surveys. One type requires people to make a phone call to record their opinion (usually an 800 or 900 number). This kind of survey often appears on television shows.

The other type features a polling group or market researcher calling random phone numbers and asking questions.

☐ A polling group asking whom you are going to vote for in the next election

☐ A 900 number to call to show whether or not you support legalized abortion

☐ A market researcher calling to find out which household products you use

3. **Personal Interview**

Surveys are often conducted in person—in a shopping mall, on a street corner, or door to door.

☐ A survey conducted in a mall asking questions about you and your shopping habits

☐ A survey conducted on a street corner about what factors influence your choice of political candidates

☐ An official door-to-door census

Out of all the preceding choices, which survey would you be most likely to participate in? Why?

____

____

You Try It

What do you think are the advantages and disadvantages of each type of survey?

1. Mail — Advantages: ____________________

   Disadvantages: ____________________

2. Telephone — Advantages: ____________________

   Disadvantages: ____________________

3. Personal Interview — Advantages: ____________________

   Disadvantages: ____________________

▶ Sample answers are on page 183.

Survey Organizations

You may already be familiar with the large survey and polling organizations in the United States. Here is a brief outline of some of them and what type of surveys they do.

Public Opinion Polls

Major public opinion polling organizations include the *Gallup* poll, the *Roper* poll, and the *Harris* poll. These organizations survey people about their opinions in a wide range of areas, including politics, consumer products, and social issues.

They are highly respected organizations because they are *independent*. They do research and surveys on their own and sell their work to the media and other organizations that might be interested in reporting on it. These polling organizations provide objective data and statistics.

Current Population Survey

Like the U.S. Census, the *Current Population Survey* (or CPS) is a survey taken by the government, but it is taken *monthly*, not every ten years as is the census.

CPS is a survey of a random sample of 100,000 Americans, and it provides us with such well-publicized information as the unemployment rate, income statistics, and the birthrate.

Newspaper and Network Polls

Most major newspapers and television networks in the United States conduct surveys of their own.

Like Gallup, Roper, and Harris, the media also provide information about public opinion that they obtained from their own surveys.

▶ **EXERCISE 6**

Pick up a copy of a city newspaper. Look for headlines and graphs that give statistics. Write down or cut out the headline, the date, the publication, the statistics, and *what polling or survey organization provided the statistics*. Keep a journal of surveys and statistics over a period of time. This will provide an interesting "year in review" at the end of the term. It will also reinforce how frequently statistics are used.

MARGIN OF ERROR

Percentage of Olson City Citizens That Have:	African-Americans	Whites
Savings Accounts	45%	75%
Checking Accounts	30	50
Stocks/Mutual Funds	10	25
Savings Bonds	10	20

Margin of Error ±2%
Source: *Olson City Journal*, 1992

What do you think *margin of error* means in the chart at left?

Margin of Error: shows by how much the statistics may vary from the actual population

Many statistics include a statement of **margin of error**. This information tells the reader that the numbers given might be slightly "off." When information from a sample is applied to a whole population, some allowance must be made for possible differences between sample and population.

For example, when you consider the margin of error, the percent of white Olson City citizens who have checking accounts is determined this way:

percent reported → 50% − 2% = 48%
→ 50% + 2% = 52%

The percent of white Olson City citizens who have checking accounts is between 48% and 52%.

■■ YOU TRY IT ▶▶

1. What percent of African-Americans in Olson City have savings bonds? ____%

2. Considering the margin of error, what is the range of African-Americans owning savings bonds?

   10% − 2% = ____%

   10% + 2% = ____%

   The range is ____% to ____%

Did you find the range to be **8% to 12%**?

More About Margin of Error

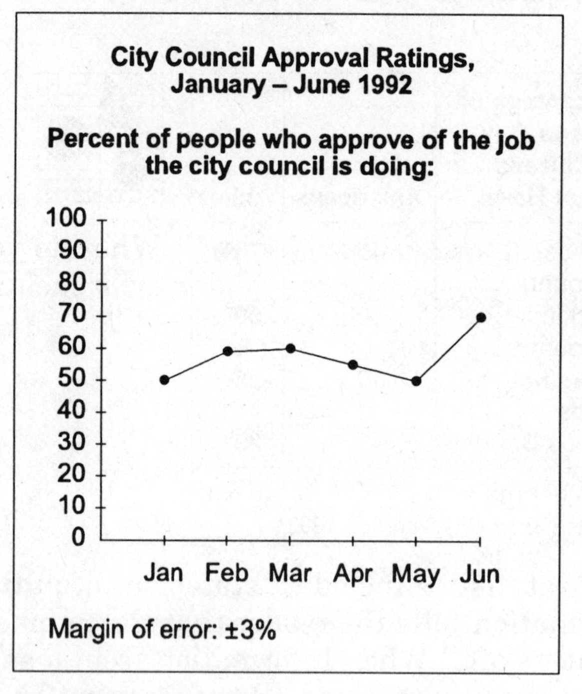

Source: *Daily Herald*

What can you say about the city council approval rating from March to April of 1992?

____

____

____

YOU TRY IT

1. What is the approval rating in March? ____%
2. What is the March approval rating adjusted for margin of error? ____% to ____%
3. What is the approval rating in April? ____%
4. What is the April approval rating adjusted for margin of error? ____% to ____%
5. Compare the low end of the range for March and the high end for April. March: ____% April: ____%
6. Is it possible that approval *rose* between the two months?
   Yes No

Answers are on page 183.

At first glance, you may decide that the approval rating went down between March and April. However, *it is possible that the approval rating for the city council actually rose a slight amount between these two months*. See the explanation in the answer key on page 183 to find out why.

Acceptable margins of error are 5% or *less*. Otherwise, the statistics are considered inaccurate. Margins of error can also be written in decimal form. For example, ±2% would be shown as ±.02.

Margins of error are a good reminder that the reporting of statistics is not an exact science.

1,500 teenagers, in a random sample from West Valley, were asked the following questions. Here's how they responded.

	Yes	No	Don't Know
"Do you have a close relationship with your parents?"	22%	71%	7%
"Are you happier now than you were five years ago?"	39	46	15
"Do you think that most of your friends are happy?"	35	55	10

Margin of error ± 4%

▶ **EXERCISE 7** Use this data to answer the following questions.

1. What is the *range* of West Valley teenagers who do not have a close relationship with their parents?
___% to ___%

2. Considering the margin of error, is it possible that more teenagers surveyed *are* happier now than five years ago? What are the ranges of possible percents?

___

___

3. How many teenagers surveyed said that most of their friends are happy? ___

4. How many teenagers surveyed answered "Don't know" to each question?

a. ___

b. ___

c. ___

▶ Answers are on page 183.

A Look at Political Polling in History

If you follow the news during election time in the United States, you are sure to be bombarded with statistics. For weeks before the election, you are showered with figures:

- Candidate A is ahead of candidate B by 10%.
- Candidate C's name is recognized by 55% of the voting population.
- Candidate D rose 5% in the polls following a persuasive speech.

And now, with our rapidly improving technology, we even get up-to-the-hour statistics *as the election is taking place*. In **exit polls**, surveyors ask voters whom they voted for as they leave the polling site, and these figures are broadcast on television.

Now that you have a good understanding of samples and surveys, you know where these numbers come from. How reliable are these figures? The statistics we get from the major *independent* polling organizations (see pages 121–122) are generally very reliable. However, this was not always the case. Let's take a look at a famous case in which a survey provided very *unreliable* results.

President Thomas E. Dewey?

Polling and sampling techniques have improved greatly over the years. In the past, pollsters relied on a method of matching the U.S. Census exactly to get a representative sample. For example, if the census reported that 12% of the population was black, pollsters made sure that 12% of their sample was black. If 23% of the population commuted into a city to work, the sample had the same percentage.

Why, then, in 1948, did all three major national polls—Gallup, Roper, and Crossley—predict that Thomas Dewey would defeat incumbent Harry Truman? Truman won the election by a wide margin.

Many studies were done after this famous mishap. Looking back, we find that some mistakes were made in the surveying and sampling methods used by the polling organizations.

1. The surveying took place weeks before the election. The surveys did not take into account a late trend favoring Truman.

2. Pollsters were allowed to survey anyone they found in one area, while other areas and people weren't represented at all.

Since this embarrassing event, a great deal has been done to improve polling techniques. Surveys are now conducted right up to the day of an election. In addition, when surveyors are sent out to take a poll, they are given specific names and addresses of people to interview. These names are gathered by a true method of random sampling.

Basically, by putting the names of *every single American of voting age* on a list and choosing 1,200 of them at random, polling organizations ensure the principle of random sampling: *each member of the population is equally likely to be chosen.*

CRITICAL THINKING

Very early in an election year, citizens find out who the "front-runners" are in a political race—based on public opinion polls. Even before we hear all of a candidate's points of view, we read about his or her "chances of winning." Do you think public opinion polls have a good or bad influence on our political system?

Consider This Situation

A voter does some research on all of the candidates for office and decides he likes how Candidate X stands on almost all issues. The voter then learns—from early public opinion polls—that Candidate X is favored by only 10% of the population. Believing that his candidate does not stand a chance of winning, the voter decides to cast his vote for Candidate Y. He does this because he is strongly against Candidate Z and he believes that Candidate Y can defeat Candidate Z.

Do you think that early public opinion polls influence voters? Do they influence the outcome of an election? Do they serve a legitimate purpose in a democracy, or do they cloud the issues? Can you come up with arguments for and against such polls?

GROUP PROJECT

Write Your Own Survey

In your first group project, you had a chance to conduct a survey that was already written for you. Now you'll have a chance to write and conduct your *own* survey.

Step 1. What do you want to find out?
Your first step as a group is to choose a topic that interests you. Is there anything that you have been wanting to know about people in your class, neighborhood, town, or city? Here's your chance to learn more.

Remember that you can ask "fact" questions such as "Do you read the newspaper on a regular basis?" or "opinion" questions such as "Do you think our congressmen and congresswomen truly represent their constitutents?" Choose a general topic to explore. It could be anything from gun control or the environment to restaurant preferences or favorite books.

Decide now as a group what you would like to learn more about from your survey and jot it down here:

____

Step 2. Choose a population, sample, and survey method.

Remember that your population will be the group that you want information about. Is it your class? your school? your neighborhood? neighborhood adults? U.S. citizens? residents of your state?

Population: ____

Of course, you will not be able to select a true scientific random sample. For the purposes of this survey, your goal should be to survey as representative a group as possible. Remember: that means that each member of the population has an equal chance of being chosen for the sample.

Decide as a group how you will choose your sample and the steps you will take to make sure your sample is random.

Now decide how you will conduct your survey. Mail is probably not feasible for you as it is time-consuming and too expensive for this purpose. How about a telephone survey? Is this a good way to survey your population? Or should you try personal interviews? Will all members of your group or just some of you survey participants? Now is the time to decide.

Survey method: ______________________________

Step 3. Writing the questionnaire.
Here are some things to keep in mind as you write your survey questions.

- Ask straightforward, easy-to-understand questions. Questions that can be answered *yes* or *no* are easiest to tabulate.
- Try not to show bias in your questions. The person being asked the questions should not be able to detect your opinion on the issues.
- Keep the questionnaire short; it should not take longer than 5 or 6 minutes to answer all the survey questions.
- "Test" your questions on group members first. Replace or change the wording of any questions that are confusing.

Sample Questionnaire Format

	Yes	No
Question: ______________________________ ______________________________	☐	☐
Question: ______________________________ ______________________________	☐	☐
Question: ______________________________ ______________________________	☐	☐
Question: ______________________________ ______________________________	☐	☐
Question: ______________________________ ______________________________	☐	☐

CHAPTER FIVE

Evaluating Data

DO STATISTICS LIE?

Used Car Dealer: "Numbers don't lie. I'm showing you the selling prices of all 8 cars sold off this lot in the past week. The car you're looking at is priced below all of them! This car is a great deal."

Customer: "How do I know these figures are true? You could be making them up."

Used Car Dealer: "Here's the printout from our computer. Check the selling prices for yourself."

Customer: "If you're telling the truth, this car is cheaper than anything you've sold lately. I'll take it."

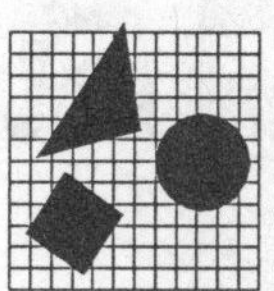

- Suppose the car dealer's numbers are accurate. What do they tell the customer about the car that's on the lot now?
- What conclusions is the customer jumping to using the data from the car dealer?

The car dealer used a list of numbers, or data, to try to sell a used car. But should the data really convince the customer? What does the purchase price of other cars tell you about the quality and price of this *particular* car?

Conclusion: a judgment or decision based on facts

The data that the dealer uses might be accurate, but the **conclusion** the customer draws from the data is *not valid.*

Accurate data and statistics don't lie, but be careful of how the numbers are interpreted.

In this chapter, you'll get a chance to see how accurate and reliable data can be misunderstood and misinterpreted.

The practice you now have in making statements about data will help you detect conclusions that are not valid.

■■ You Try It ▸▸

Here's the data shown to the customer on page 132. Suppose the customer is deciding whether or not to buy an '87 Toyota Corolla for $7,900. Make two statements about the data. One statement should compare the price of the Corolla to the prices of these other cars.

Car	Purchase Price
'90 Chevy Caprice	$10,900
'89 Ford Taurus	9,500
'87 Mazda RX7	8,900
'90 Buick Regal	11,200
'88 Ford Taurus	8,400
'90 Honda Accord	11,400
'89 Honda Accord	10,900
'88 Ford Fiesta	7,950

Statement 1: The Corolla ______________________________

____

Statement 2: ______________________________

____

Your statements may sound similar to the ones below. Based on the data, were you able to say whether the Corolla is a good buy?

1. The Corolla's price of $7,900 is lower than any other car price shown by the dealer.
2. The purchase prices of the other cars range from $7,950 to $11,400.

The data cannot help you decide whether or not the Corolla is a good buy. All you can really conclude is that the Corolla is less expensive than any of the cars sold in the past week. As a consumer, you know that there is a lot more to buying a car (especially a used car) than comparing the price to the dealer's other prices!

This example shows that it is important not only to pay attention to statistics, but also to think about *how you interpret the statistics.*

CRITICAL THINKING

Was the car dealer lying? No. He provided accurate and truthful data. But was he being honest? Explain your reasoning.

DRAWING CONCLUSIONS FROM DATA

U.S. Leads World in Prison Use, a Survey Finds

WASHINGTON—By a widening margin, the U.S. remains the world's biggest user of prisons, a private research group reported . . .

Number of People in Prison, per 100,000 Population—1990

Country	Prisoners per 100,000
United States	455
South Africa	311
Venezuela	177
Hungary	117
Canada	111
China	111

Source: *The Sentencing Project*

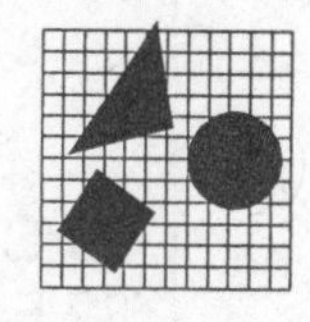

"These statistics prove what I've been saying all along," said Paul. "Our legal system in the United States is the most efficient in the world. Just look at how many criminals our system arrests, convicts, and puts behind bars, compared to these other countries."

What do you think about Paul's statements? Did he interpret the data accurately? Do these statistics prove that the U.S. legal system is the most efficient in the world?

YOU TRY IT ▶▶

Write a statement that summarizes the purpose of the data above. Remember to include information about what, when, where, and how.

Statement: ______________________________

____

____

Does your statement sound something like this?

The chart shows the number of people in prison in 1990, per 100,000 population, in six different countries.

At first, you may think Paul's statement makes sense. Maybe more prisoners do indicate more efficiency. But don't rely on just the numbers when you analyze Paul's statement! You need to use your thinking skills as well.

The data may be accurate, but the *conclusion* Paul draws from the data may not be accurate.

Asking yourself the following questions may help you draw other conclusions:

- What does it mean to have an "efficient legal system"? Does it mean that, when a crime is committed, the criminal is arrested, tried, convicted, and punished as swiftly and as inexpensively as possible?
- What information does this data give us about how many crimes are committed in the different countries and how quickly and inexpensively the criminals are imprisoned?

In fact, Paul's statement is not "proved" by the data. His statement is not necessarily untrue, but Paul needs other data to support his claim.

It is often difficult to decide whether someone's interpretation or conclusion is accurate. Always question how data is interpreted and what conclusions are being drawn.

Invalid: without foundation in fact, truth, or logic

Do not take it for granted that, if the numbers are accurate, any statement made *about* the numbers is accurate. *Your critical thinking skills are as important as the numbers.* The experience you now have in making statements about data will help you detect *invalid* conclusions.

The exercise on the following two pages will allow you to practice using your thinking skills when analyzing data.

▶ EXERCISE 1 Each set of data below is accompanied by a statement that is not supported by the data.

a. Write a short explanation telling why each interpretation is *not necessarily an accurate conclusion.*

b. Then write a statement that is supported by the graph.

Example: "Wow!" says Jeanette. "People in Watertown must get really dirty in July and August. Just look at how much soap they buy!"

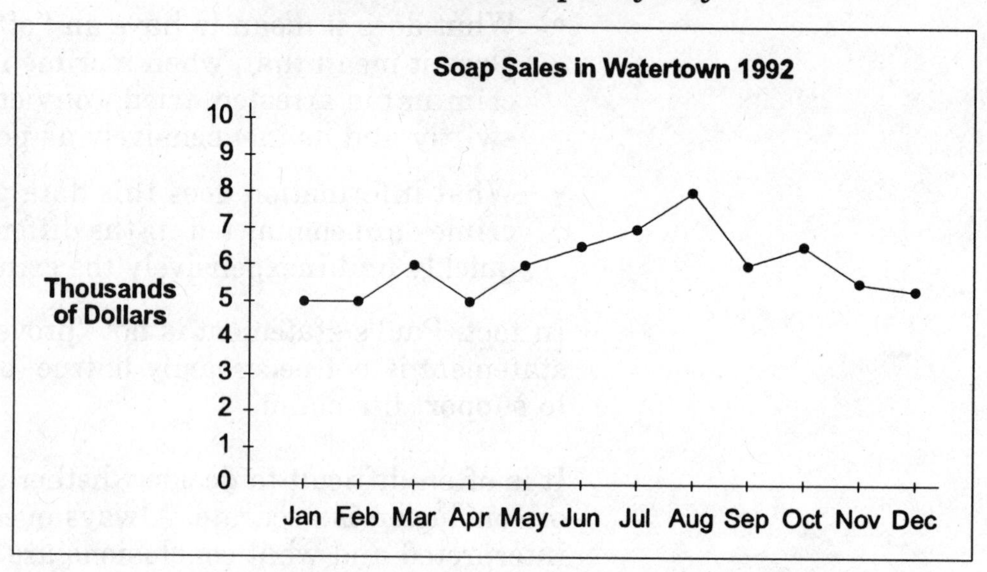

a. The fact that soap sales are high in July and August does not necessarily mean that people actually need more soap then or that they are dirtier.

b. Soap sales in Watertown were highest in August and lowest in April.

1. "Numbers don't lie. Oral contraceptives are the best method for birth control if you don't want to be sterilized."

Failure Rates of Most Common Contraceptive Methods

Sterilization—less than 1%
Oral Contraceptives—6%
Condoms—16%
Diaphragms—18%

Source: Alan Guttmacher Institute. Copyright 1991 by *USA TODAY.* Reprinted with permission.

a. ______________________________

____

b. ______________________________

____

2. "As the graph indicates, the citizens of North County pay far too much for police protection. Our police officers' salaries are the highest in five counties!"

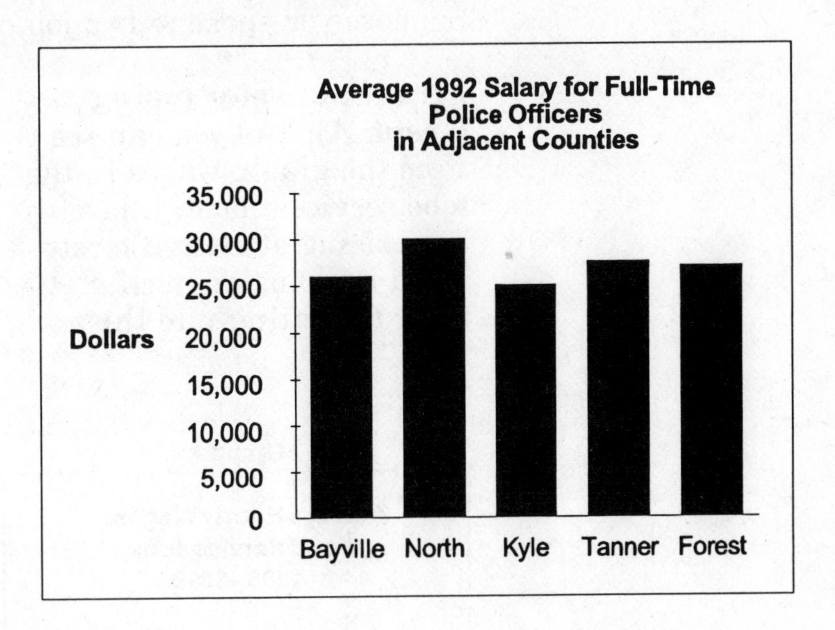

a. ____

____

b. ____

____

Fire Deaths

. . . The United States has one of the highest annual fire death rates among developed countries:

Country	1991 Deaths per Million People
South Africa	29
Canada	22
United States	21
Norway/Finland	19
Great Britain	17

Source: U.S. Fire Administration, Fire Protection Association of South Africa

3. "Look at these statistics, Bob. It just goes to show you—our fire protection is not as great as you say it is. Look at how many people die in fires in the United States compared to other countries."

a. ____

____

b. ____

____

▶ Answers are on pages 183–184.

MISINTERPRETING GRAPHS

Suppose you spoke to two job counselors.

First Counselor (using Graph A): "As you can see from the graph, wages in the food service industry have been rising at a terrific rate. You'd be doing yourself a favor by getting into this field."

Second Counselor (using Graph B): "The figures indicate that food service wages are pretty flat. You might want to look into a field where wages are rising more quickly."

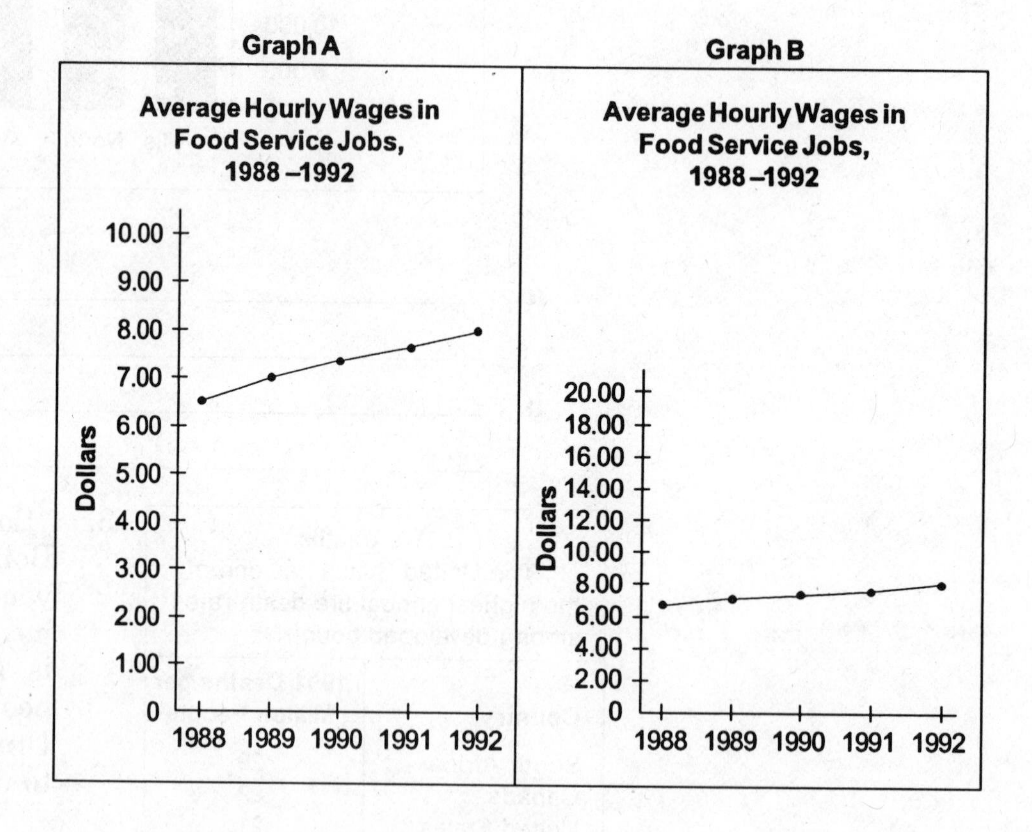

- Which person is telling you the truth? Are they both reading the data correctly?
- How can you make a decision based on two graphs that seem to be contradicting each other?

Fortunately, your work in this book and your own critical thinking skills will help you interpret graphs that might be misunderstood.

■■ YOU TRY IT ▸▸	Make a statement summarizing Graph A.	Statement: ______ ______
	Does your statement hold true for Graph B?	Yes No

Although Graph A seems to show a steep rise in wages and Graph B seems to show little rise, **both graphs actually present the same data.**

To see this clearly, fill in the blanks below. Use one graph to answer the question and the other graph to check your answer.

Between 1988 and 1992, food industry wages rose from about ______ per hour to about ______ per hour.

Both graphs show a rise from about **$6.50 to $8.00 per hour**.

Scale: the increase in values along the axis of a graph

If both graphs use the same data, why do the graphs look so different? The difference is in **scale**. As you learned in Chapter Two, the scale used to present data can vary. Sometimes the values on the scale can be multiples of 2, sometimes multiples of 5, 10, or 100.

The scale on Graph A increases by ______ at each interval.

The scale on Graph B increases by ______ at each interval.

With a smaller scale **(increases of $1.00)**, the data line on Graph A rises higher to reach the $1.50 wage increase. With a larger scale **(increases of $2.00)**, the data line does not need to rise very much at all.

In addition, compare the spacing between the dollar amounts on the scale of Graph A to the spacing of the dollar amounts on Graph B.

Which graph has larger spacing between the values on the vertical axis? ______

The spacing between the values on Graph A is larger than the spacing on Graph B. Therefore, the data line is steeper on Graph A.

So, how can you tell, by looking at the graphs, whether wages have increased a lot or a little?

You decide. You learned in Chapter One that a number is neither small nor large until you compare it to another number. Neither Graph A nor Graph B will tell you whether $1.50 is a large increase. Understanding the data, collecting more information, and *using your thinking skills* will help you decide.

■■ YOU TRY IT ▶▶

Which of the following will help you decide whether a wage increase of $1.50 per hour is large or small? Circle *Yes* or *No* for each.

- ■ the hourly wage increase in a related industry — Yes No
- ■ the number of jobs in the food industry — Yes No
- ■ the salary of a manager in the food industry — Yes No

Did you decide that it would be useful to know how wages increased in related industries? If so, your thinking skills are right on target. Although you might be interested to know how many jobs are available and what a manager's salary is, these figures will not help you evaluate the $1.50 wage increase.

Your work in this lesson should help you keep in mind two important things:

1. **When analyzing data on a graph, pay close attention to the scale.** Don't look at just the shape of the data line. Use the skills you've developed in making statements about data to determine what is happening on the graph.
2. **Use your thinking skills.** Don't rely on what someone else says about the data. Find appropriate numbers to compare with the data. Use your judgment in deciding about the size of a figure.

▶ EXERCISE 2

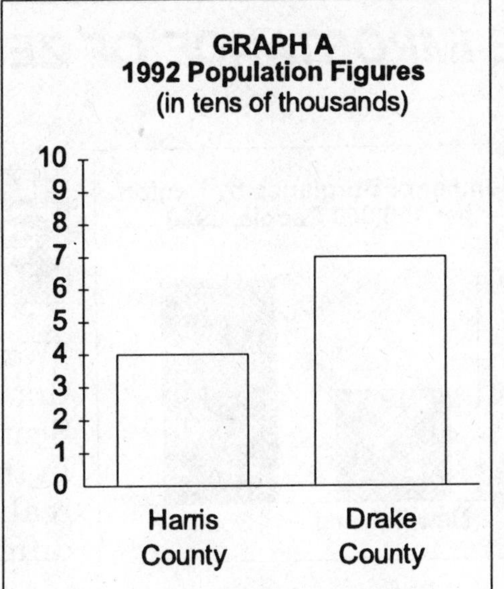

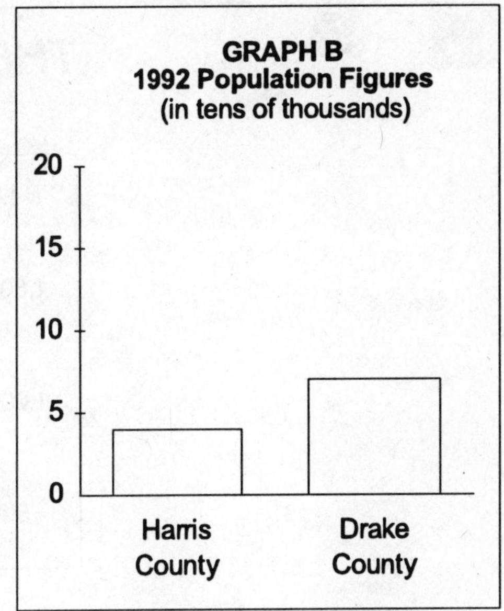

Statement 1: "The graph shows how much greater the population is in Drake County than in Harris."

Statement 2: "As you can see, there's not a huge difference between Harris and Drake in terms of population."

1. Which statement do you think was made about Graph A? Which one was made about Graph B?

a. Statement 1 refers to Graph _____.

b. Statement 2 refers to Graph _____.

2. Make a statement comparing the 1992 population of Harris County with the 1992 population of Drake County.

____

____

3. Decide whether each of the following numbers would be useful in discovering if there is a big difference in population between the two counties. Circle *Yes* or *No*.

Yes No **a.** the population figures for other counties in the state

Yes No **b.** the number of people per square mile in each county

4. Write a paragraph describing the scale in each graph and how the scale influences the height of the data bars.

▶ Answers are on page 184.

THE IMPORTANCE OF ZERO ON A SCALE

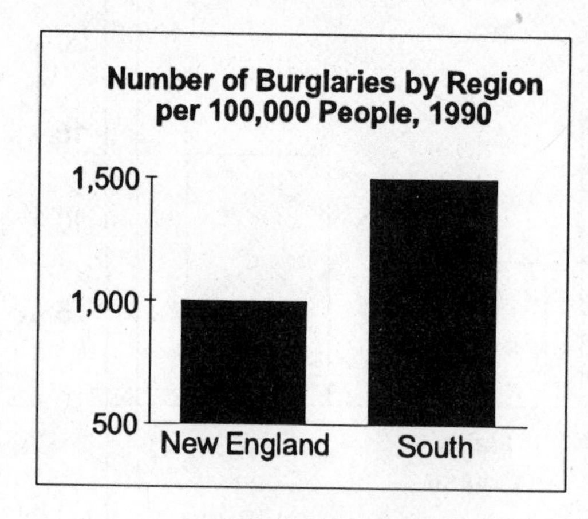

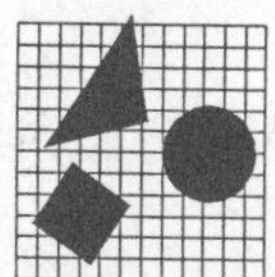

"Wow!" said Myra. "Look at this. The burglary rate in the South is about twice the rate in the Northeast. I didn't realize there was such a big difference."

Myra is correct that the bar showing the South's burglary rate *looks* about twice the height of the New England bar. But let's look at the data more carefully. Fill in the blanks below.

- In 1990, there were ________ burglaries per 100,000 people in New England.
- In 1990, there were ________ burglaries per 100,000 people in the South.

Is 1,500 twice as large as 1,000? Of course not. Then why does the graph make it appear that way?

The graph above may be misleading because it does not have a **zero** on the vertical scale. To see how important a zero is to the scale, compare the graph above to the one below—a graph that *does* include the zero.

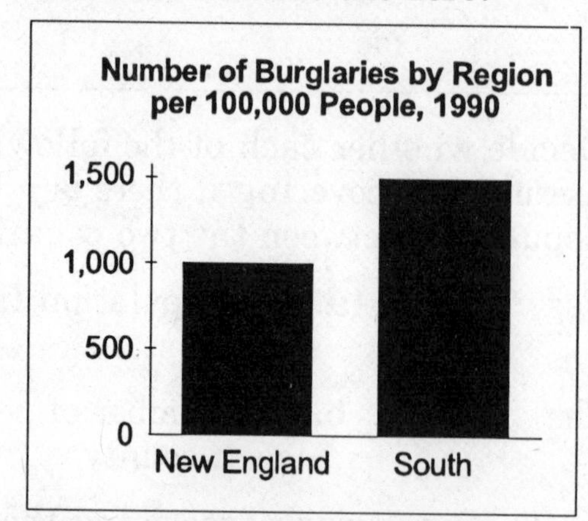

As you can see, by including the full length of each bar (including the values between 0 and 500 that were left off the graph above), the graph pictures the correct proportion between 1,000 and 1,500.

Source: *1990 Uniform Crime Reports*, FBI

It is important to look not just at the *picture* of the graph but also at the **labels** and **scale**. Beware of graphs that do not show a zero on the scale. Chances are that the proportions on the graph are misleading.

Zero Break

Sometimes you will see a graph that does show a zero on the scale, but then shows a **break** in the axis. The graph below is an example. A break like this is called a **zero break** or a **notch**.

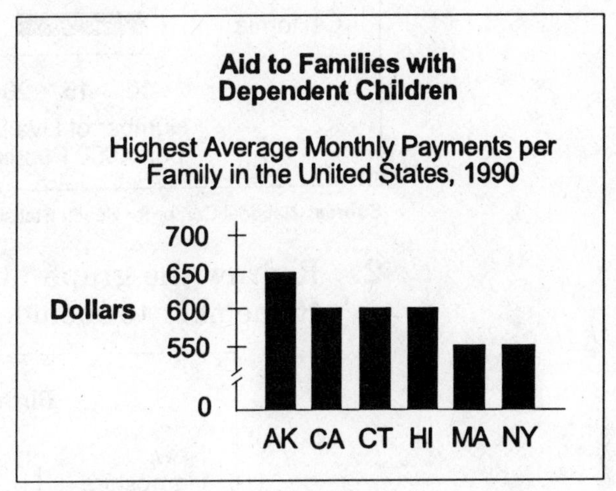

Source: U.S. Department of Health

Because all the data points to be included on the graph were above $500, the artist decided to save space and really focus on those values.

Is this graph misleading? Not to people who use their thinking skills. Whenever you see a graph that includes a zero break, remember that the scale of the graph is different than it would be if all the values were included.

A graph showing a zero break is better than a graph that shows no zero on the scale at all.

To construct a graph using a zero break:

1. Find the lowest and the highest values you will need to graph; mark these on the vertical axis.

   For example, on the graph above, the lowest value to be graphed was $550, and the highest was about $650.

2. Below the lowest value on the vertical axis, put a slash to indicate a break in the axis. Under the slash, write in a zero.

3. Finish graphing as usual.

▶ EXERCISE 3

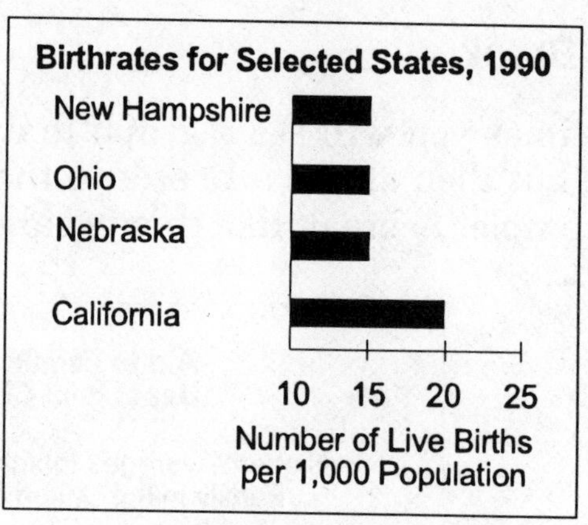

Source: National Center for Health Statistics

1. "The birthrate in New Hampshire is half the rate in California according to this graph."

   Is this an accurate reading of the data? Why or why not? ____________

   ____

   ____

   ____

2. Redraw the graph above so that it is not misleading. Remember to include a zero on the scale.

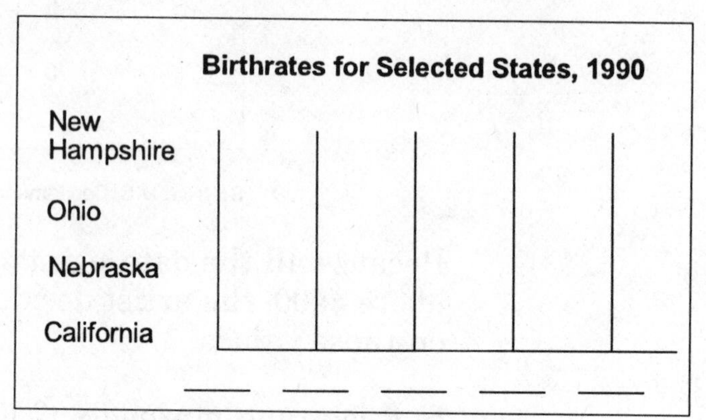

Source: National Center for Health Statistics

3. Use the data below to construct a line graph. The graph has been started for you. Use a zero break between 0 and 30.

Average Points per Game
NBA Scoring Leader
Michael Jordan

1987: 37.1
1988: 35.0
1989: 32.5
1990: 33.6
1991: 31.5
1992: 30.1

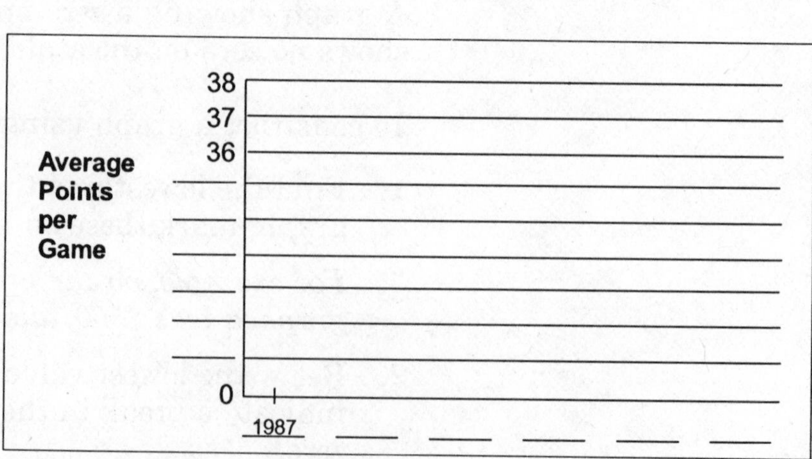

▶ Answers are on page 184.

UNDERSTANDING CORRELATION

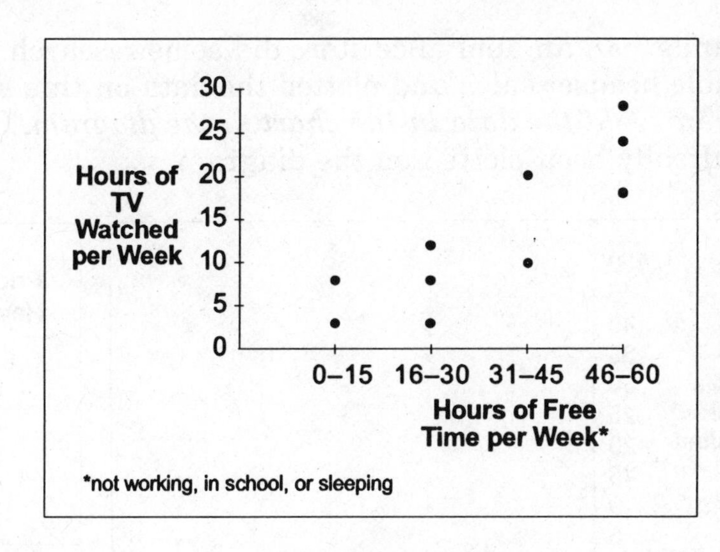

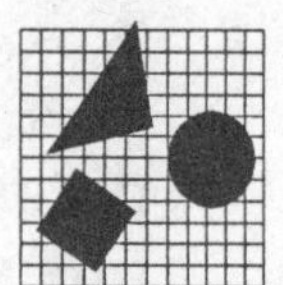

Think back to what you learned about scatter diagrams in Chapter Two. Is there a correlation between hours spent watching TV and hours of free time?

Correlation: a relationship between values

The scatter diagram above shows that there *is* a **correlation** between the two values graphed. As the number of hours of free time increases, so does the number of hours spent watching TV.

For example:
People with 25 hours of free time per week watch ________ (fewer, more)

hours of television than people with 10 hours of free time.

You should have written *more* in the space above.

The relationship shown above is called **positive correlation** because as one value rises, the other rises. To show correlation on a scatter diagram, a diagonal line is often drawn to show how most values cluster in a general pattern.

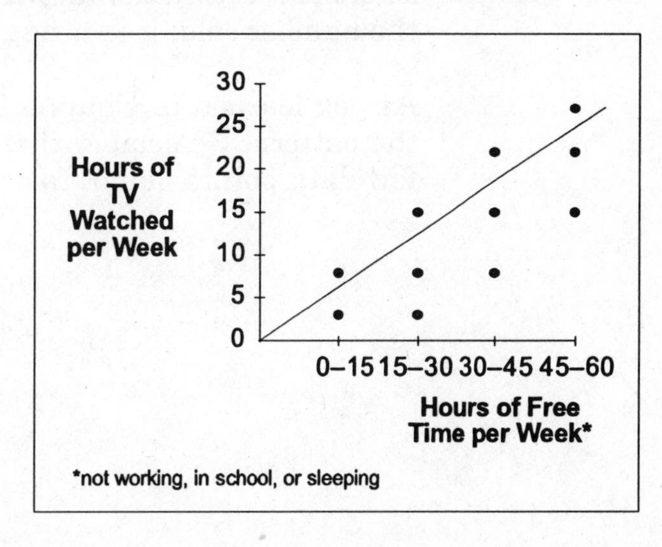

Another kind of correlation is called **negative correlation.** In this relationship, as one value rises, the other falls. Let's look at an example.

A manager in an appliance store did some research into portable headset sales and plotted the data on this scatter diagram. *Add the data in the chart to the diagram.* Other data has already been plotted on the diagram.

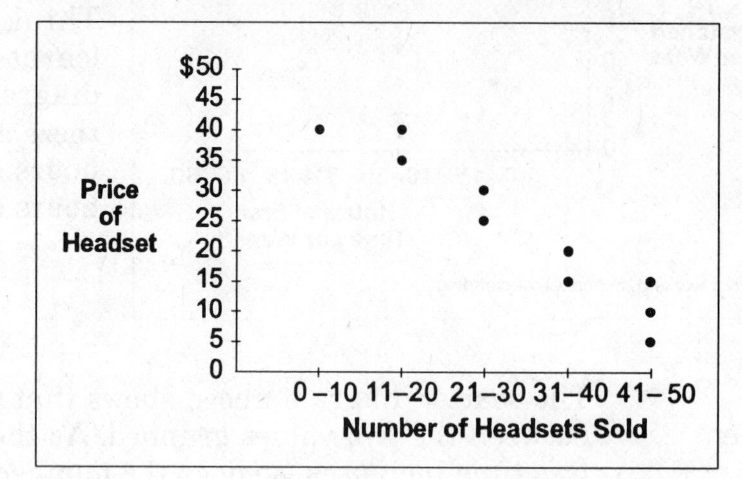

Price of Headset	Number of Headset Sold
$45	10
40	10
35	21
30	22
25	40
20	40

What is the relationship between the two values on the diagram?

As the *price of the headsets* ______________ (rose, fell), the number of headsets sold ______________ (rose, fell).

Did you write **fell** in the first blank and **rose** in the second blank? It is also true that as the price of the headsets rose, the number of headsets sold fell.

There is a correlation between the price of stereo headsets and the number sold; it is a **negative correlation.**

As you learned in Chapter Two, some values may not follow the pattern. Remember that a correlation can exist even if a few data points do not fall within the pattern.

Correlation vs. Cause

A recent consumer survey produced the data at right.

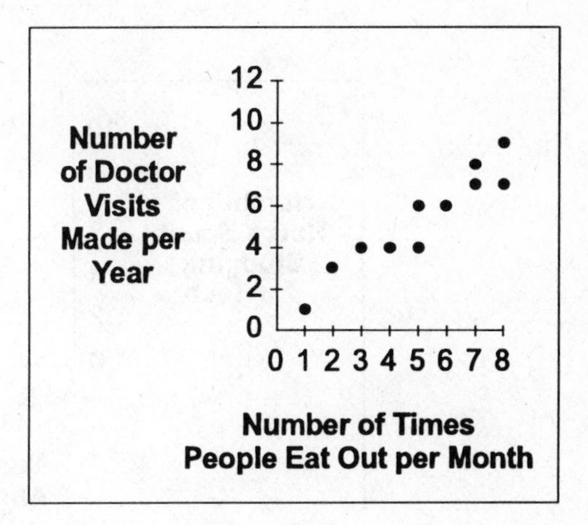

"Hey, look at this," said Kathleen. "The more often people eat out, the more often they go to see a doctor. Information like this makes you never want to eat in another restaurant!"

- Is there a correlation between the number of times people eat out and the number of doctors' visits they make?
- Which of the following do you think is the cause?
  - ☐ Frequently eating in a restaurant *causes* people to see a doctor more often.
  - ☐ More frequent doctor visits *cause* people to eat out more.
  - ☐ Neither of the above.

Actually, although there is a correlation between these two values, neither one is the **cause** of the other. Kathleen jumped to a faulty conclusion when she analyzed the data. She did not understand that *correlation* is not the same thing as *cause*.

If eating in restaurants does not cause more frequent visits to the doctor, why does this correlation exist?

In fact, trips to the doctor and trips to a restaurant increase as a third value increases—income. The more money people make, the more they eat out in restaurants and go to the doctor. People with less money tend not to seek frequent medical care or eat out often because of the expense.

Sometimes, though, when there is a correlation between two values, one *can* be the cause of the other. In other words, one value can *have an effect on* the other value.

CRITICAL THINKING

There is a negative correlation between a person's chances of being murdered and his or her income. In other words, the lower a person's income, the more likely it is that he or she will be murdered. Why do you think this is so?

■■ YOU TRY IT ▸▸

Circle the value that has an effect on the other value in each of the following correlations.

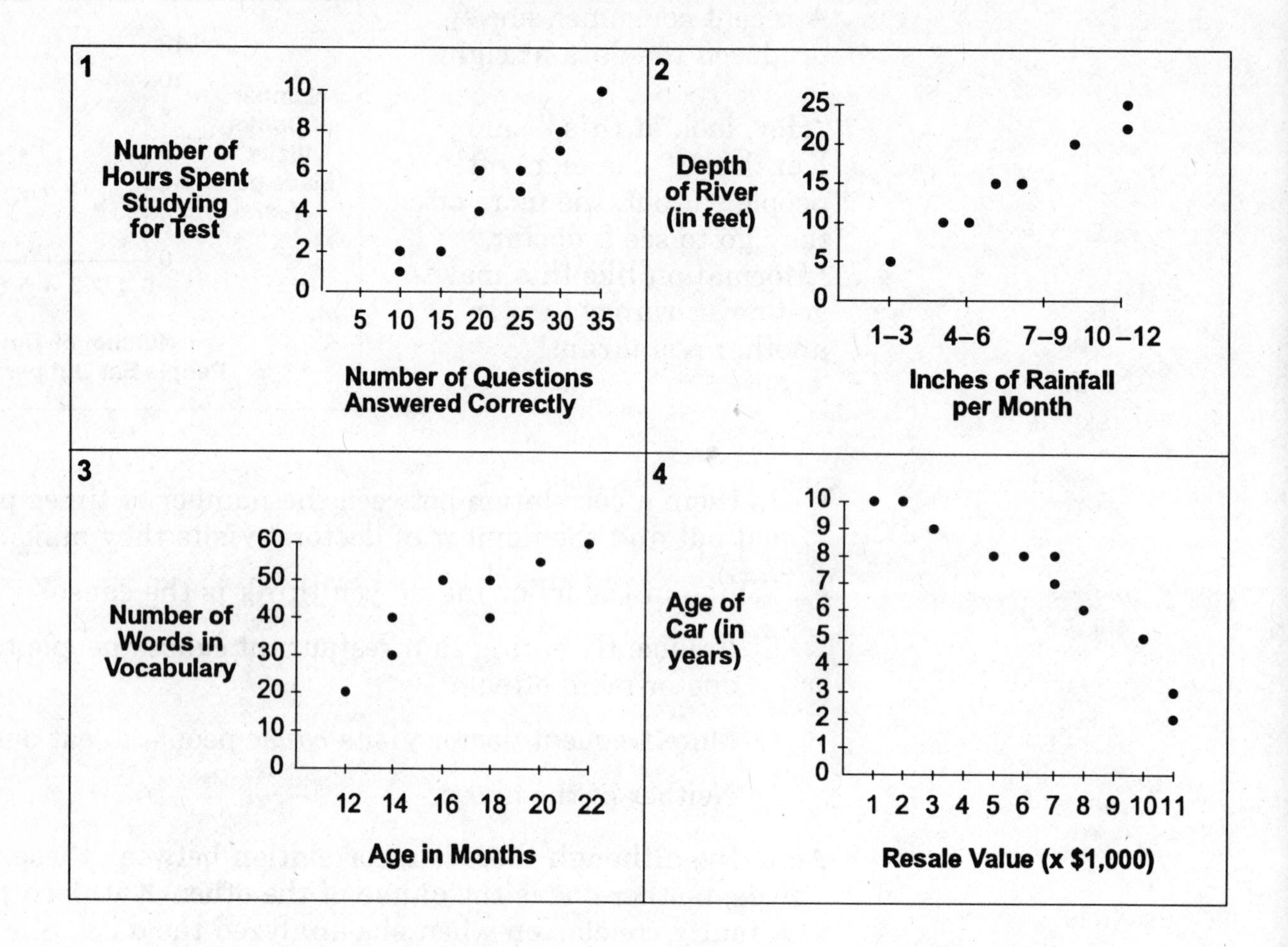

You should have found that

1. **the number of hours spent studying** had an effect on the number of right answers
2. **the amount of rainfall** had an effect on the depth of the river
3. **age** had an effect on vocabulary
4. **the age of a car** had an effect on its resale value

These four examples show a causal relationship between values. Remember, however, that *not all correlations show causation* (cause).

▶ **EXERCISE 4** Use the data below to answer the questions that follow.

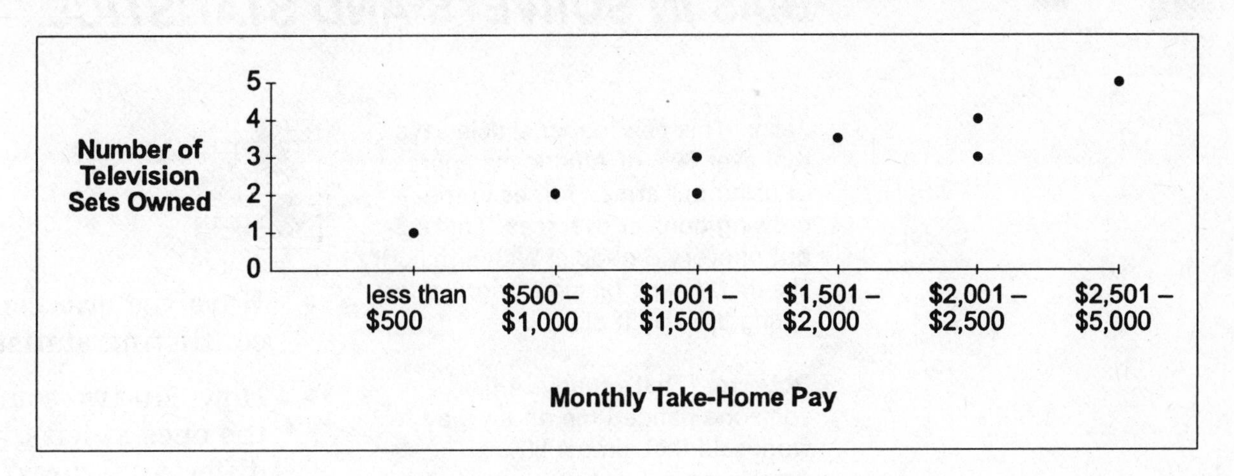

1. a. There is/is not a correlation between ________________ and ________________.

   b. "Now that I've seen these figures, I'm going to go out and buy another television set! My income is sure to go up."

   Is the analysis accurate? Why or why not?

   ____

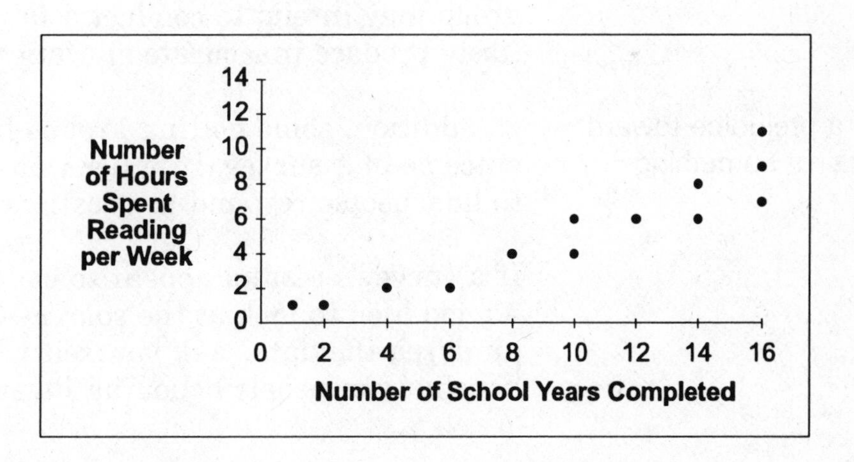

Fill in the blanks.

2. a. As the number of school years completed ________________, the number of hours spent reading per week ________________.

   b. There ________________ a correlation between school years completed and time spent reading.

▶ Answers are on page 184.

BIAS IN SURVEYS AND STATISTICS

Jean: "This newspaper article says that over 60% of Americans approve of using our armed forces in the growing conflict overseas. That's 3 out of every 5 people! With a majority like that, I can't blame Congress for voting the way it did."

Roberto: "That's funny. A flier someone handed me on my way to work said that almost 80% of Americans did *not* approve of getting involved in the war. What's going on?"

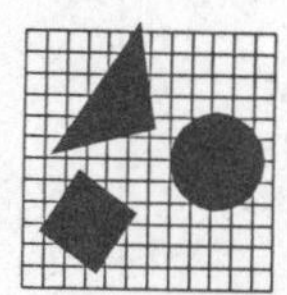

- Have you ever come across conflicting statistics?
- How can two sources like the ones at left give such differing figures?

As you learned, some survey organizations are **independent** groups; in other words, they have no financial or political interest in their survey results. In general, their methods are respected and accurate.

Some surveys, however, are not as reliable. On page 116, you saw the inaccurate results of a survey conducted with a "nonrandom" sample—the horror movie-goers. Even though a group may intend to conduct a fair survey, flawed methods likely produce inaccurate statistics.

Bias: a prejudice toward or against something

In addition, some polling groups have a **bias** concerning the outcome of a survey. In other words, they have a *preference* as to how people respond to questions on the survey.

If a survey's results appear to contradict other statistics, it is a good idea to look at the *source*—the group that gathered and analyzed the data. Ask yourself whether the source might benefit from people believing its survey results.

■■ You Try It ▸▸

Suppose that the figures on the flier Roberto received were compiled by one of the groups at right. Put a check mark next to the group you think provided the statistics.

☐ People for Peace, an organization that lobbies heavily against armed force involvement

☐ Make America #1, a group that promotes the use of U.S. military power in conflicts worldwide

People for Peace would probably benefit if we believed that almost 80% of Americans were against getting involved in the war. These statistics "support" its cause. Make America #1 would probably have no interest in reporting that 80% of Americans did not support its point of view.

Of course, People for Peace may be reporting valid and accurate statistics. We simply cannot tell just by looking at them. However, if a source shows the possibility of bias, it is a good idea to find statistics from other sources to compare.

CRITICAL THINKING

Suppose the statistics on Roberto's flier came from an independent polling agency. Would you be more likely to believe them? Why?

▸ EXERCISE 5

Put a check mark next to each survey conducted that you think might show *bias*. Ask yourself if the people conducting the survey might have a preference for what the survey results showed.

☐ **1.** A group called Citizens for Smith conducts a survey to find out how many people plan to vote for Smith in the city council election.

☐ **2.** A soft-drink company reports a survey revealing which brand of soft drink teenagers prefer.

☐ **3.** Americans Tough on Criminals surveys crime victims on their views concerning capital punishment.

☐ **4.** A newspaper polls local communities to find out how they feel about gun control.

☐ **5.** An inventor, trying to get financial support for her idea, does a survey asking people if they would buy a computerized laundry basket.

☐ **6.** A television network conducts an exit poll outside a voting center to project the winners in an election.

▸ Answers are on pages 184–185.

Biased Methods

How might an organization go about gathering statistics that support its point of view? To be sure of getting the "right" responses, the polling group could use several different methods. Let's take a look at a couple of them.

Sample Bias

To obtain the results it wants, a group could survey only certain people—people who are likely to have the same point of view as the survey group. You saw an example of sample bias in Chapter Four in the survey of moviegoers. Sample bias is sometimes done intentionally and at other times by mistake.

As you do the activity below, remember that Make America #1 is the organization that promotes the use of military power.

YOU TRY IT

Which of the groups at right would Make America #1 most likely want to survey to find support for its cause?	☐ a community with a large military base ☐ a university town known for its antiwar demonstrations

You probably know that **the first community** would be the best choice for Make America #1. Although not *everyone* in the first community would be in favor of military involvement, there would certainly be a higher percent there than in the second group!

This example demonstrates the importance of a *random sample.* Almost any organization can find and survey a group that is known to support its positions. However, the statistics obtained would not necessarily reflect the opinion of most people.

Statistics are often accompanied by an explanation of the random sample and how it was obtained. If you suspect that the statistics are not based on a random sample, you should consider the possibility of bias and look for other sources and studies on the same topic before drawing conclusions.

Question Bias

Another way that survey groups can affect the outcome of their polls is by asking leading, or "loaded," questions. This is called **question bias**. The wording of this type of question actually *directs* the respondent to answer in a certain way.

Go back and read the dialogue between Jean and Roberto. Both statistics have something to do with "military involvement," but are they really contradicting each other?

Suppose the 60% quoted in Jean's article answered yes to the following question:

Do you approve of using our armed forces in the growing conflict overseas as a peacekeeping force to protect the innocent citizens of both countries?

And suppose that the 80% quoted in Roberto's flier answered no to this question:

Do you approve of getting involved in the war overseas by sending in armed American troops to fight in the battles of other countries?

There is a big difference in the numbers quoted because there is a big difference in the questions asked.

In addition, many survey questions are designed so that the person answering the question must choose from a multiple-choice format. But what happens if a person's true opinion is not listed as one of the choices? Many times, the answers recorded in a survey are not an accurate representation of people's true opinions.

CRITICAL THINKING

Answer yes or no to the following questions:

Are taxes a good thing? ☐ Yes ☐ No

Is America a free country? ☐ Yes ☐ No

Are Americans hard-working people? ☐ Yes ☐ No

You may have found yourself wanting to answer "not always" or "yes, but . . ." Can you see that yes/no type questions often do not allow you to express your true opinion?

■■ You Try It ▶▶

The two organizations below surveyed people concerning their opinions on the U.S. space program.

Each group asked a "loaded" question to get a response that matched its own point of view. Match the organization to the question you think it would ask.

Space Is Our Future supports increased spending for space exploration.	"Are you tired of our government spending millions of dollars on useless efforts in space exploration while our own air, water, and plant life on Earth is in jeopardy?"
Home Earth seeks to divert government funds from space exploration toward environmental programs here on Earth.	"Do you believe that our government should put more resources into vital technological advancements such as space exploration so as to maintain our scientific superiority in the world?"

Did you match **Space is Our Future with the second question** and **Home Earth with the first question?**

Can you see how it would be difficult to answer no to either one of these questions? The wording of each question leads us to answer yes. The survey statistics obtained from the responses to these two questions would be very different, wouldn't they? By using biased questions, each group could claim support for its own position.

▶ EXERCISE 6

Decide if each of the following survey situations is biased or unbiased. Write a brief explanation telling why.

1. To find out how city residents felt about a new factory being built in the midst of a large neighborhood, the chairman of Citizens for Economic Development asked, "Do you agree that new jobs and more taxpaying industry are important developments in this area of economic frailty?"

   biased or unbiased

   Why? ____________________

   ____

   ____

2. A television network surveyed people to find out what percent planned to vote in an upcoming presidential election.

biased or unbiased

Why? ____

____

____

3. To find out how city residents felt about a new factory being built in the midst of a large neighborhood, a group called Save Our Homes asked, "Are you willing to sacrifice your homes and community to a factory that will destroy the environment?"

biased or unbiased

Why? ____

____

____

4. A researcher surveyed people outside an exclusive shopping mall to find out how Americans were affected by the current recession.

biased or unbiased

Why? ____

____

____

5. A consumer opinion researcher stopped people coming out of an ice cream parlor and asked them what their favorite dessert was.

biased or unbiased

Why? ____

____

____

▶ Answers are on page 185.

CRITICAL THINKING WITH DATA

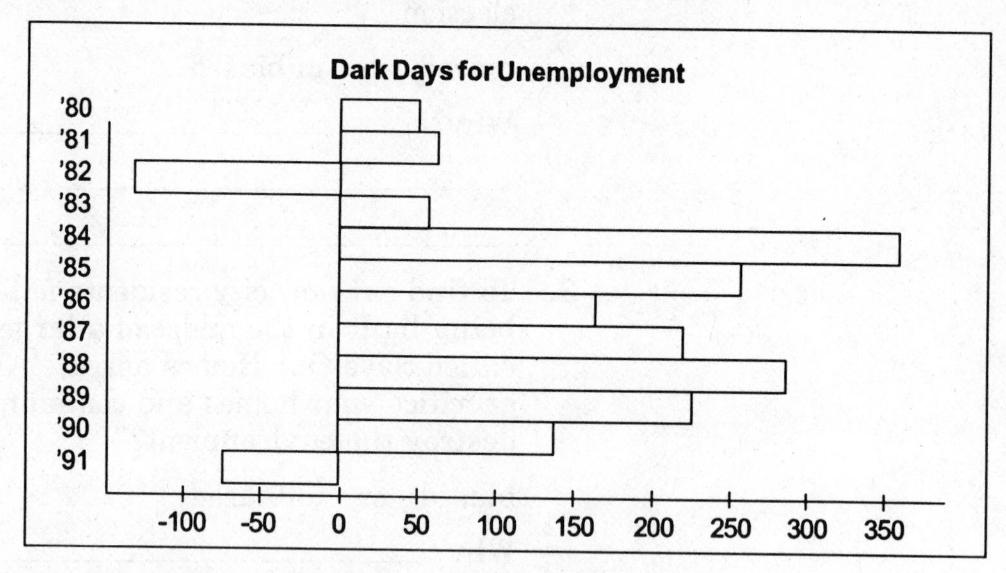

Source: Copyright 1991 by *USA TODAY.* Reprinted with permission.

Put a check mark next to the statement(s) that are true based on the graph.

☐ **Statement 1:** In 1991, an average of 73,000 jobs were lost monthly due to cutbacks in the defense industry.

☐ **Statement 2:** There was tremendous job growth in the United States during the 1980s.

☐ **Statement 3:** During the 1980s, the U.S. economy was strong and healthy.

☐ **Statement 4:** Between 1980 and 1990, 1982 was the only year in which the average monthly change in jobs was a job loss.

Critical Thinking: careful evaluation and judgment

An analysis or interpretation often accompanies data. Don't *assume* the analysis is accurate—use your **critical thinking** skills to see if the analysis is supported by the data.

At first, each statement above might appear true based on the graph. Let's find out why **only statement 4 is actually based on the data provided**.

For each statement, think about these questions:

- Is anything missing?
- Can the data be interpreted differently?

Statement 1: In 1991, an average of 73,000 jobs were lost monthly due to cutbacks in the defense industry.

Is anything missing? Yes. The graph gives no information about the causes of job loss. We don't know if cuts in defense were the cause.

Remember: even if you believe the statement to be true, the question asks which statements *are supported by the graph*, not which ones are true based on your prior knowledge or opinion.

■■ YOU TRY IT ▶▶

Statement 2: There was tremendous job growth in the United States during the 1980s.

Can the data be interpreted differently? How? Hint: What does *tremendous* mean? Based on the graph, is there any way to tell how large the 1980s job growth was?

____

____

____

The graph shows clearly that there was job growth in the 1980s—more jobs were added than were lost. But how can you decide whether this growth was "tremendous"?

In Chapter One you learned to compare numbers. You learned that a number is neither small nor large *until you compare it to another number*. In the example, what is the 1980 job growth being compared to? Job growth in the 1970s? Job growth in Japan?

■■ YOU TRY IT ▶▶

Statement 3: During the 1980s, the U.S. economy was strong and healthy.

Is anything missing? Can the data be interpreted differently?

____

____

A strong and healthy economy depends on more than job growth. To allow you to make a statement like this, the graph would need to show data concerning the inflation rate, the unemployment rate, gross national product, and other data.

Statement 4 is based on only the data provided on the graph.

■■ You Try It ▶▶

Let's look at another example:

Put a check mark next to the statement(s) that are true based on the graph.

☐ **Statement 1:** In 1989, blacks living in Queens, New York, were as wealthy as whites.

☐ **Statement 2:** In 1989, median income for both black and white households in Queens, New York, was about $33,000.

☐ **Statement 3:** Due to the success of affirmative action, blacks in Queens are finally nearing economic equality with whites.

☐ **Statement 4:** The rising median income among black households in Queens, New York, is an example of the progress being made by African-Americans nationwide.

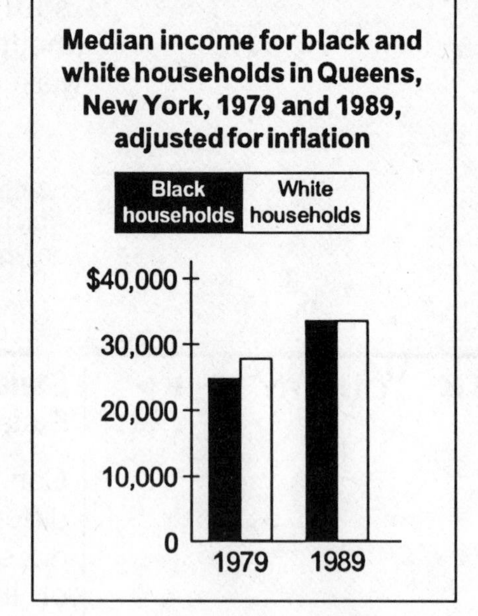

Source: U.S. Bureau of the Census

Let's look at the statements more closely.

Statement 2 is the only statment that can be made based on the graph.

Statement 1 is too general. Income is only one indicator of wealth. We do not know the savings, real estate holdings, or other assets of the households shown.

Statement 3 cannot be made based on the graph because the graph gives no *reasons* for an increasing median household income among blacks.

What happens in one city does not necessarily reflect what is happening in other cities across the country. **Statement 4** cannot be made based on the graph.

▶ EXERCISE 7

Answer the questions using the graph below.

1. Make a statement about the number of undocumented immigrant women stopped by the San Diego Sector in 1991.

____

____

____

____

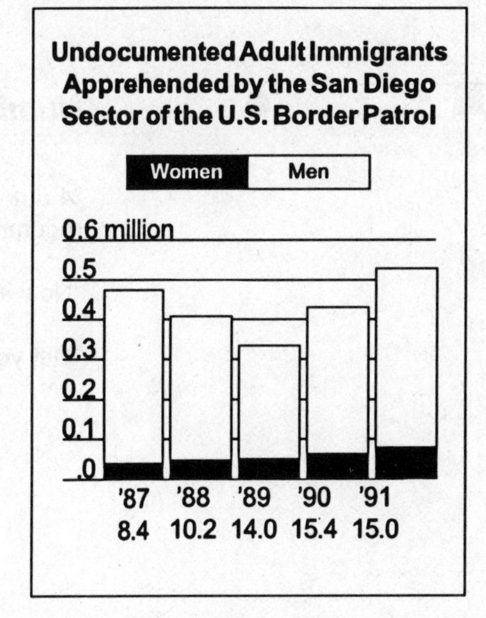

Source: U.S. Border Patrol

2. Which of the following statements can be made based on the graph?

   a. Between 1987 and 1991, the number of undocumented immigrant women stopped by the San Diego Sector increased.

   b. More women than men were stopped by the San Diego Sector in 1989.

   c. In 1991, approximately 50,000 women were stopped by the San Diego Sector as they tried to enter the United States illegally with their husbands.

   d. Between 1987 and 1991, the number of wives accompanying their husbands across the Mexico–United States border has increased.

▶ Answers are on page 185.

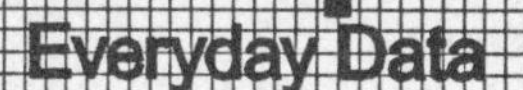

Numbers in Advertising

"4 out of 5 doctors surveyed recommend brand X!"

"Now a hot dog that's 80% fat-free!"

"Get your laundry up to 30% cleaner!"

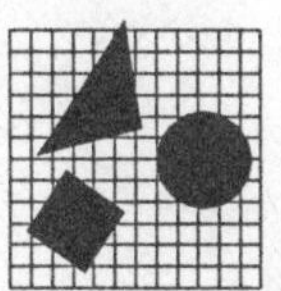

- Have you ever noticed how many advertisements use statistical claims like the ones at left?
- Are these numbers facts, and should they persuade us to buy these products?

The skills you have been working on throughout this book also apply to analyzing data found in advertisements and commercials. For each of the claims above, let's look at the numbers and *what they really mean.*

"4 out of 5 doctors surveyed recommend brand X!"

STATS

Did you know that . . . ? The government has strict guidelines for food labeling. On a food product's label, the ingredients must be listed in order of quantity. So beware of products in which sugar is the first ingredient listed!

Ask yourself these questions:

- Can you tell how large a survey was conducted? ______
- Does the advertisement tell you what question was asked of the doctors in the survey? ______

The answers should make you question the claims being made.

Few, if any, advertisements will tell how the survey was conducted. Certainly, the makers of brand X would like you to believe that they surveyed thousands of doctors in a random sample and asked them what their favorite brand of pain reliever was. More likely, however, here is how they got their numbers:

They probably surveyed a small number of doctors, possibly doctors who work in their own research labs.

They most likely asked the doctors, "Would you recommend FeelGood pain reliever to your patients?" The doctors respond yes, but the advertisement does not say that they recommend ten other brands as well.

In fact, there is nothing "untruthful" in the advertisement. However, don't be fooled into assuming too much. When you see an advertisement like this one, think of not only what is being said but also what *isn't* being said. Remember what you have learned about source and bias!

"Now a hot dog that's 80% fat-free!"

Many people see a label that says "80% fat-free" and think that a product is low in fat. But what it's really saying is that the product is 20% fat. That's $\frac{1}{5}$ of what you're about to eat.

Whenever you see a percent on a label or in an advertisement, try to restate the data so it's more specific and more meaningful.

"Get your laundry up to 30% cleaner!"

Ask yourself, "30% cleaner than *what*?"

By not identifying what a product is being compared to, and by using impressive statistics, advertisements like this one can mislead. Consumer beware!

CRITICAL THINKING

What is a good way to decide what kind of medication to buy? Can advertisements really help you make a decision?

Choose three or four advertisements that use statistics to help sell a product or service. They can be television commercials or newspaper or magazine ads.

Write down the exact statement that uses statistics. Then write down one or two reasons why the statistics do not help you decide whether to buy the product advertised.

GROUP PROJECT

A Closer Look at Surveys

Part One

Take a look at this survey description and questionnaire. Discuss the survey with your group. Use the information you have gained in this book to help you decide whether or not it is a fair survey.

Now look at the analysis on page 163. Can your group design a better survey and rewrite the questions?

In September of 1992, Conservatives for America announced the results of a poll of Americans nationwide. A sample of 500 people were asked their opinions about the political process in the United States. Survey participants were chosen randomly from a list of homeowners in the greater Washington, D.C., area.

Question: Do you think the liberal media unfairly influence voter opinion? Yes No

Question: Do you agree with most people that special-interest groups have too much power in government? Yes No

Question: Given the huge increase in government social program spending over the past year, do you feel that it is time for citizens to start paying their own bills? Yes No

Question: Do you make large contributions to any political organization? Yes No

The date a survey is made can often influence its outcome. For example, September 1992 was only a short time away from a presidential election. People's opinions might be different then than in a nonelection year, for example.

In September of 1992, Conservatives for America announced the results of a poll of Americans nationwide. A sample of 500 people were asked their opinions about the political process in the United States. Survey participants were chosen randomly from a list of homeowners in the greater Washington, D.C., area.

Do you think this group might have an interest in the outcome of the survey?

Are homeowners a representative sample of Americans?

Question: Do you think the liberal media unfairly influence voter opinion? Yes No

Adjectives and adverbs can influence responses.

Beware of questions that influence the respondent to be included in a group; many people are unwilling to go against a majority.

Question: Do you agree with most people that special-interest groups have too much power in government? Yes No

Question: Given the huge increase in government social program spending over the past year, do you feel that it is time for citizens to start paying their own bills? Yes No

Questions like this one precede the actual question with a negative comment, thereby leading the respondent to a particular response.

Questions should be specific; some people may think a $10 contribution is large, whereas others might feel that anything under $100 is small.

Question: Do you make large contributions to any political organization? Yes No

Part Two

Now go back to the survey your group wrote on pages 128–129.

Check your questions for any evidence of bias, using the sample survey as guidance. Then rewrite your questionnaire.

POST-TEST

Answer all of the questions as carefully as you can.

Problems 1–3 refer to the following data.

Davco, Incorporated			
Department	Number of Employees Who Carpool to Work	Number of Employees Who Take Public Transportation	Total Number of Employees in Department
Sales	26	55	110
Service	10	10	25
Administrative Pool	15	12	30
Maintenance	10	6	18
Management	1	5	20

1. Which of the departments below has the higher *rate* of employees who carpool to work?

(1) sales

(2) service

2. According to the chart, about 30% of Davco employees carpool to work. What numbers can you compare with this figure in order to decide if 30% is a large number or a small number?

(1) the total number of employees

(2) the percent of people who carpool to other nearby industries

(3) the number of cars in the company parking lot

3. Write a statement about the number of management employees who carpool to work compared to the number of maintenance workers who carpool.

____

____

Problems 4-7 refer to the following data.

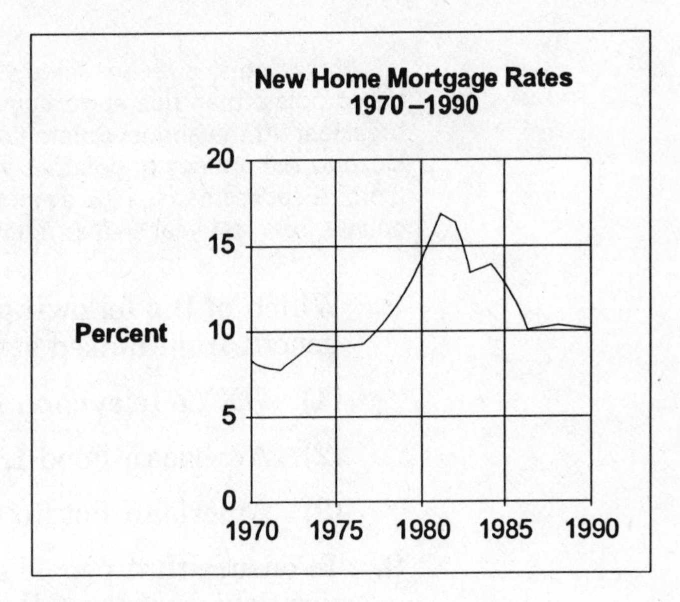

4. Which of the following best summarizes the graph?

(1) The graph shows new home mortgage rates in the United States from 1970 to 1990 in percent per year.

(2) The graph shows the amount paid in mortgages in the United States from 1970 to 1990.

(3) The graph shows new home mortgage rates in the United States.

5. What is the range of percents provided on this graph?

(1) 10% to 20%

(2) 0% to 10%

(3) 0% to 20%

6. Use the data to make a statement comparing new home mortgage rates for two different years.

____

____

7. Which statement about the graph is *not* true?

(1) Selling prices of new homes were higher in 1980 than in 1983.

(2) New home mortgage rates rose between 1975 and 1980.

(3) The new home mortgage rate in 1990 was 10%.

Problems 8–11 refer to the following newspaper article.

> ... According to a recent survey, over 70% of Americans would rather eat a baked potato than rice at mealtimes, and 20% actually eat potatoes for breakfast. At the annual Potato Lovers Picnic in Daisy County last year, close to 750 pounds of potatoes were consumed by the 3,000 participants there. A spokesman for the event reported that potato sales are up 5% in the county from last year's sales figure of $940,000.

8. Which of the following organizations would most likely benefit from biased statistics in the survey above?

(1) WXYZ television news

(2) American Food Industry, Inc.

(3) American Potato Growers, Inc.

9. To ensure that people say that they prefer potatoes to rice, which of the following questions might the survey group ask?

(1) Would you prefer a hot, buttered baked potato or a bowl of rice?

(2) Which do you prefer to eat: a baked potato or rice?

(3) Which do you have more often at mealtimes: a baked potato or rice?

10. Which of the following would be a biased sample on which to conduct the potato-or-rice survey?

(1) 1,000 people working in a city government building

(2) 500 people leaving a professional baseball game

(3) 3,000 participants at the Potato Lovers Picnic

11. According to the survey, what are the chances that an American will eat potatoes for breakfast?

(1) 1 in 20

(2) 1 in 10

(3) 1 in 5

Problems 12–14 refer to the following data.

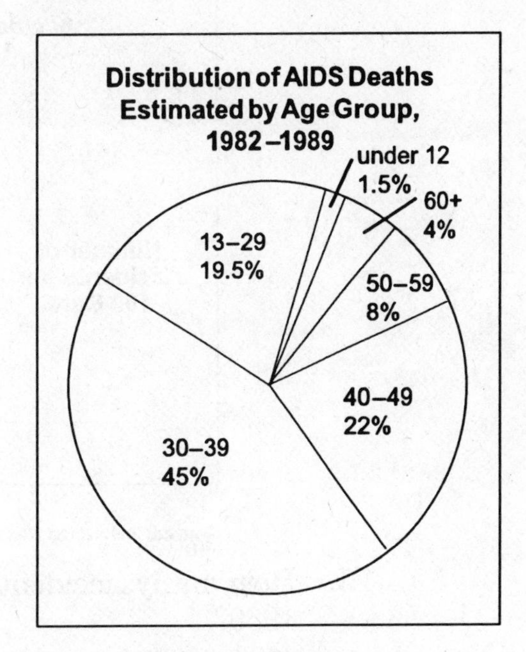

Source: U.S. Bureau of the Census

12. Which age group had the greatest percent of AIDS deaths between 1982 and 1989?

(1) 13–29

(2) 30–39

(3) 60+

13. What is the purpose of the graph above?

(1) to show how many people died from AIDS from 1982 to 1989

(2) to show the distribution of AIDS deaths, by age group, from 1982 to 1989

(3) to show estimated numbers of AIDS deaths in six age groups from 1982 to 1989

14. Write a statement about children under 12 and AIDS deaths.

____

____

Problems 15–18 refer to the following data.

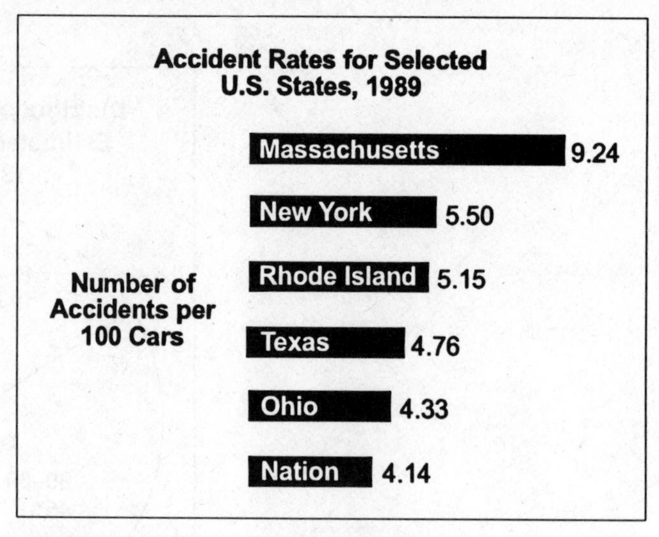

Source: Insurance Research Council

15. How many accidents per 100 cars occurred in Texas in 1989?

(1) 4

(2) 4.76

(3) 476

16. Which of the following statements summarizes the data?

(1) The chart shows the accident rate per 100 cars for the United States in 1989.

(2) The chart shows the number of accidents that occurred in 1989 in 5 states and the nation.

(3) The chart shows the accident rate per 100 cars for 5 states and the nation in 1989.

17. Write a statement that compares the accident rate of one state to the national accident rate.

____

____

18. Which of the following statements is *not* true according to the chart?

(1) In 1989, Massachusetts had more car accidents than any other state listed.

(2) New York had fewer accidents per 100 cars in 1989 than Massachusetts.

(3) Of the states shown, Ohio had the fewest accidents per 100 cars in 1989.

Problems 19–22 refer to the following data.

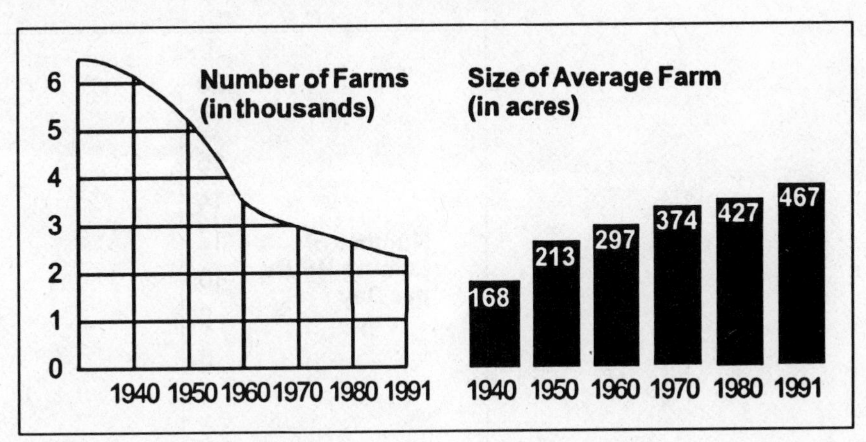

Source: U.S. Agriculture Department

19. How many acres did the average farm contain in 1950?

(1) 5,000

(2) 213

(3) 200

20. Which of the following statements best summarizes the data of both graphs?

(1) The number of farms decreased between 1940 and 1991 in the United States.

(2) The size of the average farm in the United States ranged from 168 acres in 1940 to 467 acres in 1991.

(3) As the number of U.S. farms fell from 1940 to 1991, the size of the average farm grew.

21. Write a statement about the number of farms and the average size of a farm in 1970.

____

____

22. Use data from both graphs to *estimate* how many acres of land were farmed in 1991. Then choose the answer below that comes closest.

(1) 1,000,000

(2) 2,000

(3) 500

Problems 23–25 refer to the following data.

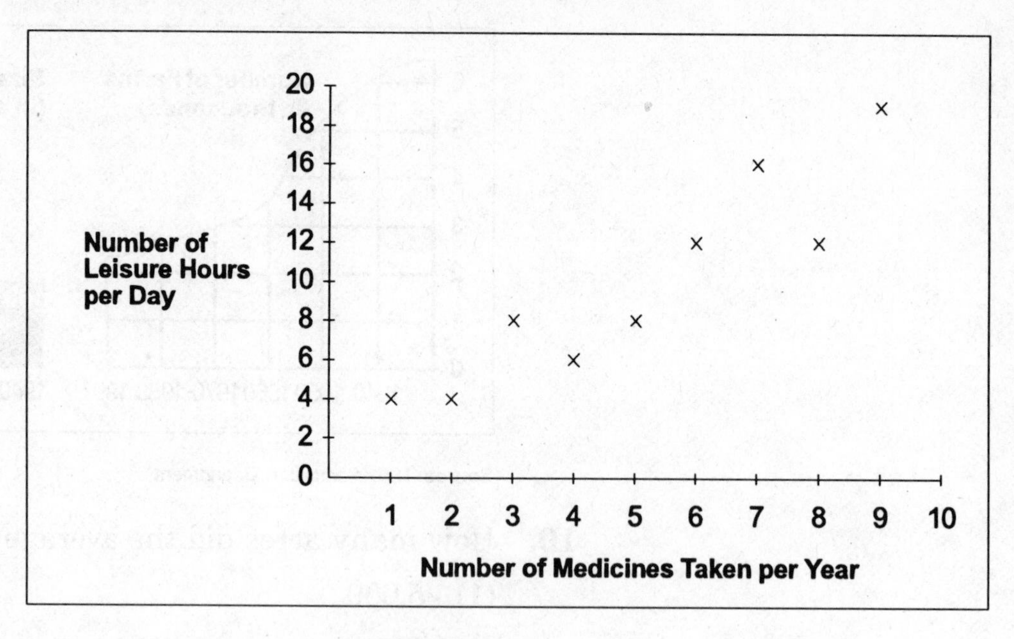

23. Is there a correlation between number of leisure hours and number of medicines taken?

(1) yes (2) no

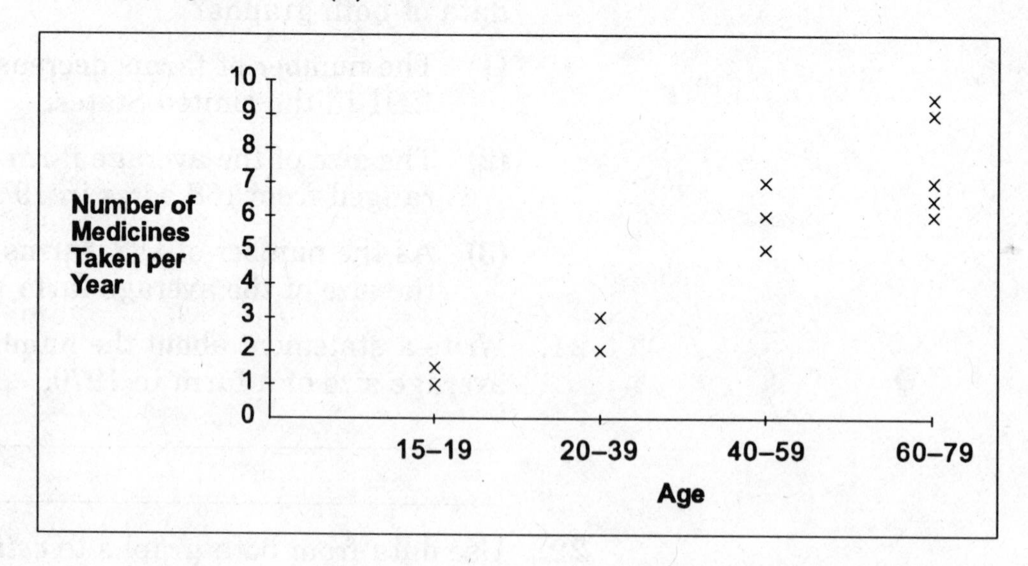

24. Is there a correlation between age and number of medicines taken?

(1) yes (2) no

25. Write a short summary of the data above. Do you think any of the three factors listed (age, leisure time, number of medicines taken) has an effect on the other factors?

____

____

Problems 26–28 refer to the following data.

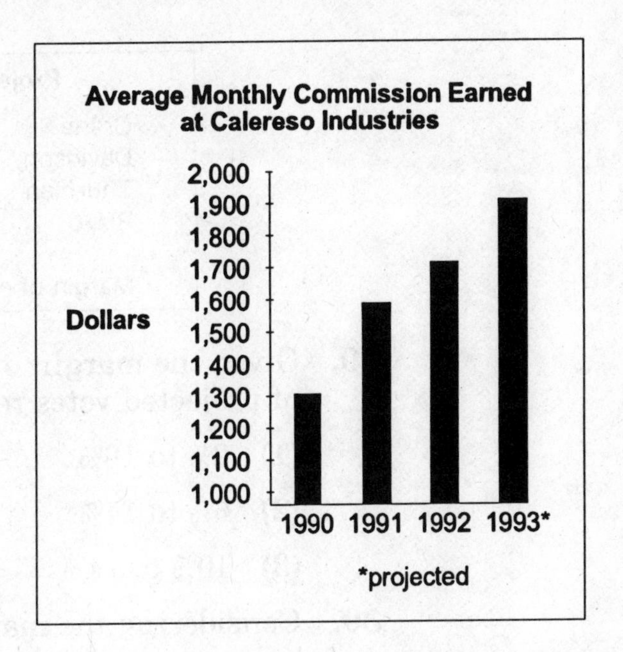

26. Write a statement about average monthly commission at Calereso Industries in 1993.

____

____

27. Which of the following statements is true according to the graph?

(1) The average monthly commission at Calereso will exceed $2,000 in 1994.

(2) The average monthly commission at Calereso was twice as high in 1991 as in 1990.

(3) The average monthly commission at Calereso rose steadily between 1990 and 1992.

28. Was the average monthly commission earned in 1992 twice the amount earned in 1990? Why does it appear that way on the graph?

____

____

____

Problems 29–32 refer to the following data.

Projected Election Results	
Crane	46%
Davidson	39%
Thurman	10%
Rizzo	5%
Margin of error $\pm$ 4%	

29. Given the margin of error, what is the range of percent of projected votes received by Thurman?

(1) 2% to 18%

(2) 6% to 14%

(3) 10% to 14%

30. Considering the margin of error, is it possible that Crane will end up with half of the votes?

(1) yes

(2) no

31. If 400,000 votes were cast for these four candidates, *approximately* how many were cast for Crane?

(1) 200,000

(2) 100,000

(3) 10,000

32. Given the margin of error in this poll, is it possible that Davidson will win the election?

(1) yes

(2) no

Problems 33–35 refer to the following data.

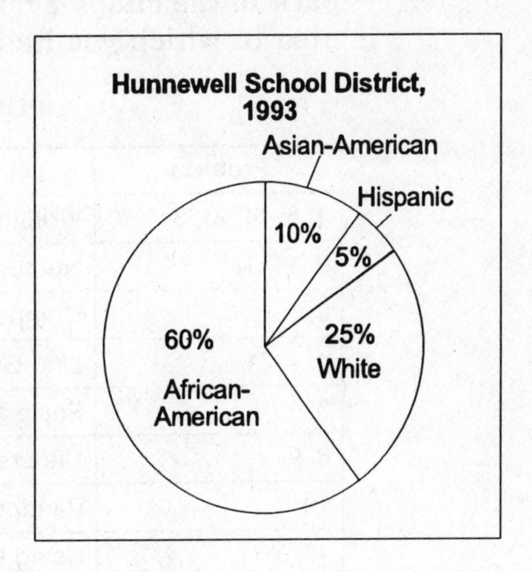

33. What are the chances that a child in the Hunnewell School District is African-American?

(1) $\frac{1}{60}$

(2) $\frac{3}{5}$

(3) $\frac{1}{6}$

34. Write a statement comparing two races in the Hunnewell School District in 1993.

____

____

35. If there are 12,400 students in the Hunnewell District in 1993, how many are Hispanic?

(1) 620

(2) 1,240

(3) 6,200

▶ Answers are on pages 185–186.

Circle any problem number that was answered *incorrectly*. Go back to the chapter number indicated and review the skill area in which you had difficulty.

POST-TEST SKILL ANALYSIS

Problem	Skill	Chapter
1, 6, 14, 21, 34	Writing Statements About Data	1
2, 14, 17	Numbers vs. Rate	1
3	Comparing Numbers	1
4, 7, 21	Line Graphs	2
5	Scale and Range	2
8, 9	Understanding Bias	5
10	Random Sample	4
11, 33	Using Data to Find Probability	4
12, 13, 35	Circle Graphs	2
15, 18	Visual Statistics	2
16, 22	Using Only the Information Given	3
19, 20	Using More than One Data Source	3
22, 31	Estimation	3
23, 24, 25	Understanding Correlation	5
26, 27	Seeing Trends/Making Predictions	3
28	Zero on a Scale	5
29, 30, 32	Margin of Error	4

ANSWER KEY

PRE-TEST

1. **(2)** California. 30,000,000 is the largest number on the chart.

2. The table above shows the 1990 **population** to the nearest million for five states.

3. Answers may vary. A sample statement is given below.

The 1990 population of Arkansas was about half the population of Alabama.

4. **(3)** 26,000,000

$$\begin{array}{rl} 30{,}000{,}000 & \text{(California)} \\ -\ 4{,}000{,}000 & \text{(Alabama)} \\ \hline 26{,}000{,}000 & \end{array}$$

5. **(1)** 40%

6. being caught in traffic

7. Answers may vary. A sample statement is given below.

About 50% of the people surveyed get angry at slow restaurant service.

8. 48%

65% + 65% + 50% + 40% + 20% = 240%

240% ÷ 5 = **48%**

9. **(1)** 35%

10. **(3)** twice

1991: about 70%

1986: about 35%
70 is two times, or twice, 35.

11. **(3)** more than 70%

The graph shows the percent rising, so the percent in 2000 will probably be larger than the 70% in 1991.

12. **(2)** People rented more videotapes in 1990 than in 1989.

Although more households had VCRs, you cannot assume that more videotapes were rented.

13.

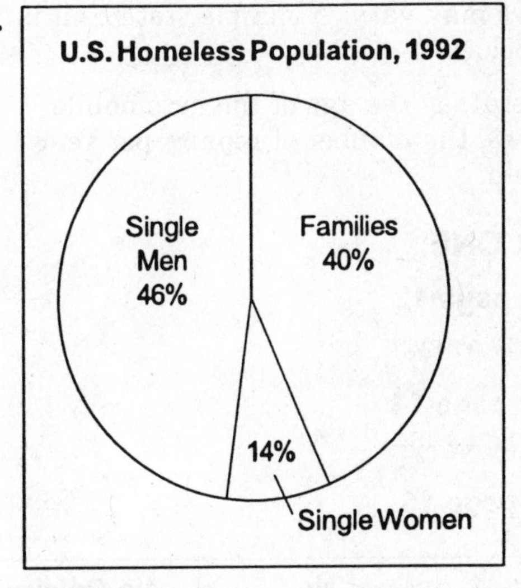

Source: U.S. Department of Housing and Urban Development

14. **(1)** 168

14% of 1,200 is the same as 14% × 1,200, or 168.

15. Answers may vary. A sample statement is given below.
More than three times as many single men as single women were homeless in the United States in 1992.

16. **(3)** 3 in 5

$\frac{3}{5}$ (2, 6, and 10 are multiples of 2)
(5 cards in all)

17. **(1)** 0

Since there are no fives, the probability of choosing a 5 is 0.

18.

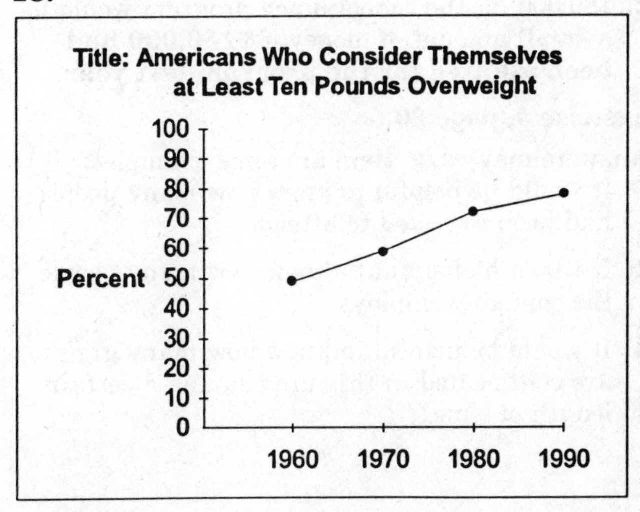

19. (1) yes

In general, as the age of a car increases, so does the number of repairs per year.

20. Answers may vary. A sample statement is given below.

In general, as the age of the automobile increases, the number of repairs per year increases.

CHAPTER ONE

Exercise 1, page 12

Answers may vary.

Exercise 2, page 14

Answers may vary.

You Try It, page 16

Yes	No	No Opinion
////	//	/

Exercise 3, page 17

1. 95%; $\frac{95}{100}$; 95:100 or 19:20
2. 8%; $\frac{8}{100}$; 8:100 or 2:25
3. 70%; $\frac{7}{10}$; 7:10 or 70:100
4. 30%; $\frac{3}{10}$; 60:200 or 30:100 or 3:10
5. 1%; $\frac{1}{100}$; 1:100
6. 80%; $\frac{4}{5}$; 4:5 or 80:100

You Try It, page 19

Answers may vary. Sample statements are given below.

1. $80,000 for the school lunch program would be a lot of money if **the total school budget was $100,000.**
2. $80,000 for the school lunch program would be a small amount of money if **$150,000 had been allotted for the program last year.**

Exercise 4, page 20

Answers may vary. Here are some examples.

1. It would be helpful to know how many people had been *expected* to attend.
2. It would be helpful to know how many people the company employs.
3. It would be helpful to know how many guns are confiscated in this area during a certain length of time.
4. It would be helpful to know how much money the artist usually gets for a painting.
5. It would be helpful to know how many years old the car is or how many miles it has on it.

Exercise 5, page 21

1. nine hundred thousand
2. eight million
3. ninety million
4. 100,000,000
5. 2,000,000,000
6. 700,000

Exercise 6, page 23

1. 1,000
2. 10
3. 10
4. 100
5. 10

You Try It, page 25

Highway

B	.001	=	.1	%
D	.0033	=	.33	%
G	.0015	=	.15	%
HH	.0015	=	.15	%
MM	.002125	=	.21	%

Exercise 7, pages 26–27

Part One

Answers may vary. Sample statements are given below.

1. Kramer, Inc., had the smallest number of satisfied employees, thirteen.
2. Walton Company and Psytech had the largest numbers of satisfied employees.
3. Psytech, with 11.5%, had the lowest satisfaction rate.
4. Horton & Co. had the highest satisfaction rate, almost 25%.
5. Although Psytech had the largest number of satisfied employees, its employee satisfaction rate was the lowest.

Part Two

Answers may vary. Sample statements are given below.

1. The unemployment rate increased from 5.4% in 1990 to 7% in 1992.

2. Teenagers have a much higher unemployment rate than the total work force.

3. In 1990, the black unemployment rate was more than twice that of white Americans.

Exercise 8, page 29

1. New York has the highest official population count.

2. Los Angeles has the highest percent of people who weren't counted.

CHAPTER TWO

Exercise 1

Part One, page 35

1. 1
2. 74
3. 152
4. T
5. F
6. T

Answers may vary. Sample statements are given below.

7. A one-ounce serving of peanuts **contains four more grams of fat than one ounce of potato chips.**

8. Although corn chips have more calories than potato chips, they contain less fat.

Part Two, page 36

Percent of Americans Holding Credit Cards

	$50,000 or More	**$25,000–$34,999**	**Less Than $15,000**
Bank Credit Card	94	**73**	**36**
Gasoline Credit Card	**54**	**33**	**19**
American Express Gold Card	**14**	**3**	**1**

Source: *Money* magazine

Part Three, page 37

1. 73
2. gasoline
3. $25,000–$34,999
4. 40
5. F
6. T
7. F
8. T

Answers may vary. Sample statements are given below.

9. The bank credit card **had the highest percent of holders in every category.**

10. Americans earning less than $15,000 had the lowest percent of card holders.

You Try It, page 39

1. 20%
2. 18%
3. 13%
4. 6%

You Try It, page 40

1. Park and Recreation Land by County
2. County
3. Acres (in hundreds)
4. 0 to 20
5. 4

Answers may vary. Sample statements are given below.

6. Dawton County has more park and recreation land than **Kane County.**

7. Kane County **has less park and recreation land than the other counties shown.**

8. San Pedro County has the most park and recreation land of all the counties shown.

Exercise 2, page 41

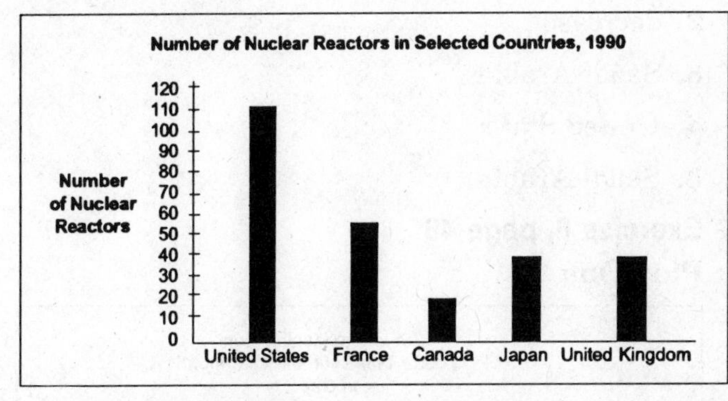

Source: International Atomic Energy Agency

Answers may vary. Sample statements are given below.

1. In 1990, the United States had **more nuclear reactors than any of the other countries shown.**

2. Canada had **the fewest nuclear reactors of the selected countries in 1990.**

3. Japan and the United Kingdom **had the same number of nuclear reactors in 1990.**

Exercise 3, page 43

1. T
2. T
3. F
4. T

Exercise 4, page 45

Part One

1. 30 years old
2. 5 years
3. above

Part Two

Answers may vary. Sample statements are given below.

1. The median U.S. age was the same in 1980 as it was in **1950**.
2. The median age in the United States rose steadily between 1820 and 1950.
3. The median age in the United States in 1990 was about 33 years old.

You Try It, page 47

1. — — — — —
2. 10
3. 3
4. fell

Exercise 5, page 47

1. risen
2. decrease
3. Saudi Arabia
4. United States
5. Saudi Arabia

Exercise 6, page 48

Part One

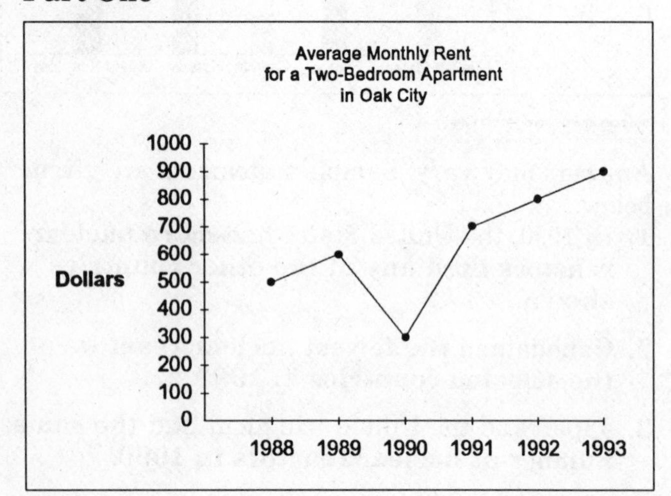

Part Two

1. A line graph shows the change over time better than a bar graph.

   Answers may vary. Sample statements are given below.

2. Rental prices in Oak City were about $600 higher in 1993 than in 1990.
3. Except for a sharp drop in 1990, rental prices rose between 1988 and 1993.

Exercise 7, page 49

Answers may vary. Sample statements are given below.

1. Between 1970 and 1990, the annual per capita rice consumption in the United States **more than doubled**.
2. Between 1985 and 1990, the annual per capita rice consumption in the United States rose by 7.5 pounds.

Exercise 8, pages 50–51

1. $100,00
2. 1986
3. 1988, $225,000
4. yes
5. $137,500

Exercise 9, page 51

Part One

NOTE: If you chose a different value in your key, your graph will be different from this one. Be sure that you have included the proper number of symbols on your graph.

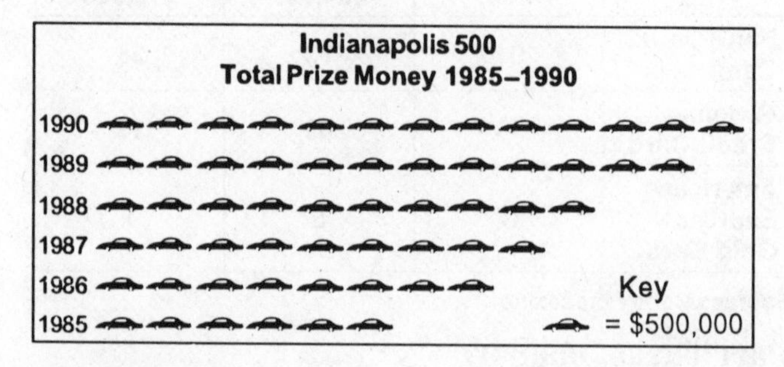

Part Two

1-2. Answers may vary.

You Try It, page 53

Answers may vary. Sample statements are given below.

1. There are fewer students from **the Middle East than from Asia**.
2. Students from Asia accounted for more than half of foreign students.

You Try It, page 53

Answers may vary. A sample statement is given below.
National defense receives about twice as much as states and localities.

Exercise 10, page 54

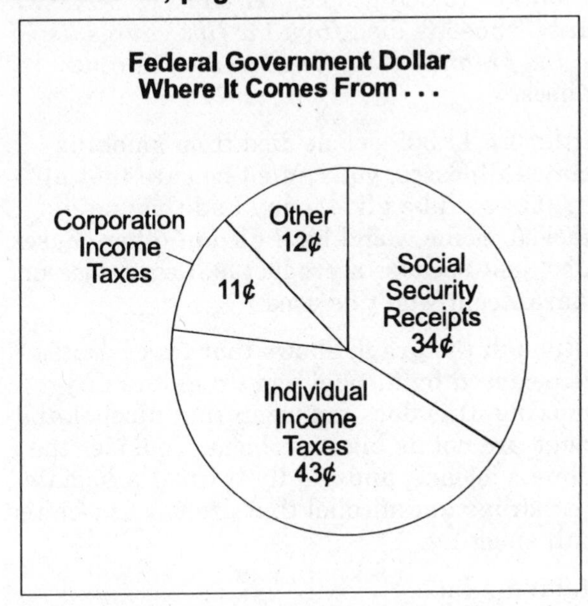

Source: Office of Management and Budget

You Try It, page 55

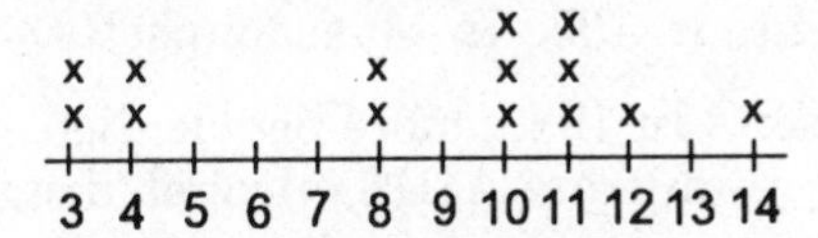

Exercise 11, pages 56–57

Part One

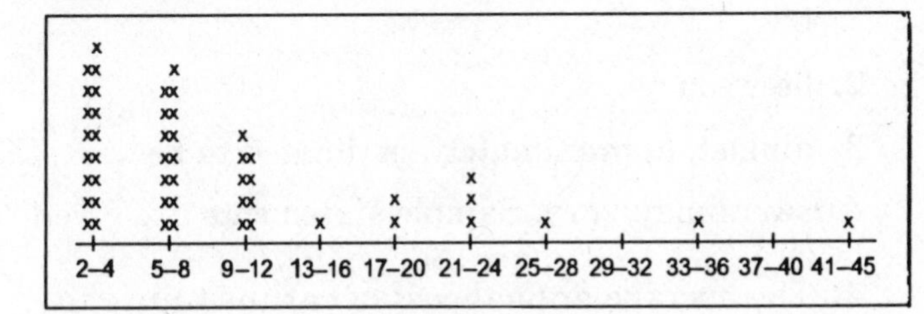

Part Two

1. Two representatives is the smallest number from any state. Twelve states have two representatives.
2. California, with 45 representatives, has the largest number. No other state has close to that many representatives.
3. There is a cluster of values around 2-12 representatives.
4. Answers may vary. Two sample statements are given below.
   a. Many states have **fairly small populations, and therefore only 2 to 4 representatives.**
   b. California has the largest population, and New York has the second largest.

Exercise 12, page 59

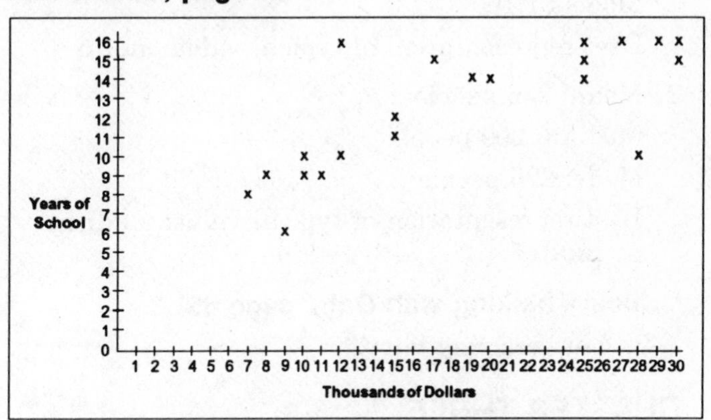

1. Yes, the higher the number of school years completed, the greater the salary.
2. Answers may vary. A sample statement is given below.

   In general, the more school people complete, the more money they earn.

You Try It, page 61

2. 24
3. $11,710 ÷ 24 = $487.92

Exercise 13, page 61

1. 4 miles
2. $70
3. $32

Exercise 14, page 62

1. $12.25
2. 6′10″

You Try It, page 63

1. 1,700: 2
   1,650: 1
   1,600: 1
   290: 2
   270: 2
   260: 2
   250: 9
   240: 3
   230: 1
   220: 1
2. 250

Exercise 15, page 63

1. $.49
2. 10 weeks

Exercise 16, page 64

1. Mean: $646.67
   Median: $575
   Mode: $1,100
   Best representation of typical value: **mean**
2. Mean: 256 people
   Median: 200 people
   Mode: 225 people
   Best representation of typical value: **median or mode**

Critical Thinking with Data, page 65

1-2. Answers may vary.

CHAPTER THREE

Exercise 1, pages 75–76

1. **No**
   The *what* is inaccurate; the graph shows the best-paying **outdoor** jobs.
2. **No**
   The *where* is inaccurate; the graph shows the price of milk at **Crescent Farm Dairy**.
3. **No**
   The *what* and *how* are inaccurate; the chart shows Americans who **believe** it's all right to lie, **in percent**, not numbers.
4. Yes

Exercise 2, pages 78–79

Part One

1. The statement does not take into account the label × *100* on the vertical axis; the number of passengers carried is 1,200, not 12.
2. The statement does not include the necessary time: 1992.
3. The statement does not tell what the 600 refers to; Carling Coach carried 600 passengers in 1992.

Part Two

Answers may vary. Sample statements are given below.

1. According to the County Taxi Association, Carling Coach carried half as many passengers as Yellow Cab in 1992.
2. In 1992, At Your Service Cab Company carried 200 fewer passengers than Yellow Cab, according to the County Taxi Association.

Exercise 3, page 82

Part One

1. A total of 13,585 people over 35 in Massachusetts died *from the five causes listed on the graph*, but many more died of other causes.
2. Although 11,305 people died from smoking-related illnesses, you cannot be sure that all 11,305 would be alive today had they not smoked. Some would have died of other causes. Also, just because a product is illegal does not guarantee it won't be used.
3. Although the graph shows that fewer deaths were caused by alcohol and drugs than by smoking, this does not mean that alcohol and drugs are not as big a problem. Consider the crime, violence, and family turmoil associated with drugs and alcohol that are not associated with smoking.

Part Two

Answers may vary. Sample statements are given below.

1. According to the Bureau of Health Statistics, 737 people aged 35 and over died in motor vehicle accidents in 1988 in Massachusetts.
2. In Massachusetts in 1988, more people died from smoking than from AIDS, alcohol/drugs, motor vehicles, and homicide/suicide combined.

Exercise 4, page 84

1. less than, almost, approximately, estimated to be
2. between
3. almost, approximately, estimated to be

Answers may vary. Sample statements are given below.

4. The average annual cost of car insurance in the United States increased steadily between 1975 and 1990.
5. The average annual cost of car insurance in the United States was about the same in 1980 and 1985.

Exercise 5, page 87

Part One

Answers may vary. Sample words and phrases are given below.

1. In 1989, immigration from Italy was **about** 3,000.
2. Immigration from Italy in 1989 was **greater than** the immigration from Spain.
3. The number of immigrants from Sweden in 1989 was approximately **the same as** the immigration from Spain.
4. In 1989, **approximately 14,000** more people immigrated from Poland than from Sweden.

Part Two

Answers will vary.

You Try It, page 89

1. Graph 2
2. Both
3. Neither
4. Graph 1

Exercise 6, page 90

1. a. Chart
   b. $\frac{1}{10}$
   $\frac{140}{1,400} = \frac{1}{10}$
2. a. Both
   b. 16
   20% of 80 = 16
3. a. Graph
   b. 210
   15% of 1,400 = 210
4. a. Neither; the chart and graph give no information about which sexes participate in which activities.

Why Change the Form of Data?, pages 92-93

2. Answers may vary.

3.

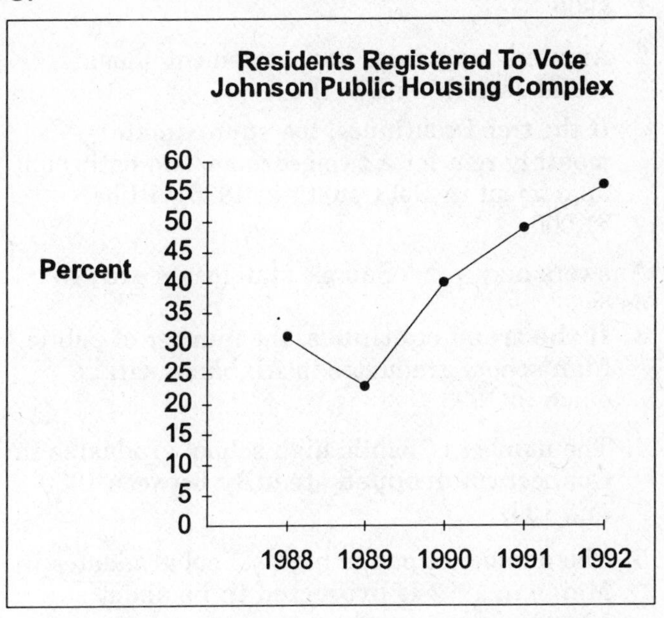

4. a. The **percent** of registered voters **rose** between 1991 and 1992, even though the first graph showed the **number** of registered voters falling.
   b. Answers may vary.

You Try It, page 94

Steps 1-3.

Northeast	South	Midwest
1,140,000	1,043,000	639,000

Exercise 7, page 95

1. **Step 1.** 540
   Step 2. Medicare: 140 ÷ 540 = about 26%
   BC/BS: 120 ÷ 540 = about 22%
   Private: 200 ÷ 540 = about 37%
   No insurance: 80 ÷ 540 = about 15%
   Step 3.

Insurance	Medicare	Blue Cross/ Blue Shield	Private	No Insurance
Percent of Patients	26	22	37	15

2. a-b.

Sun County Homeless Population	
Area	Number of Homeless People
Downtown Area	3,515
Central Square	2,755
East Side	2,090
West Side	950
Outlying	190

Exercise 8, pages 98–99

1. \$900
2. Answers may vary. Your statement should sound something like this:

   If the trend continues, the approximate monthly rent for a two-bedroom/two-bathroom apartment in Del County in 1995 will be \$1,000.

Answers may vary. Sample statements are given below.

3. **If the trend continues,** the number of public high school graduates in Alabama will go down in 1993.
4. The number of public high school graduates in Connecticut **dropped steadily** between 1980 and 1991.
5. The number of public high school graduates in Maine in 1992 **is projected to be** about 12,000.
6. Connecticut had **approximately 15,000 more** public high school graduates than Maine in 1992.

CHAPTER FOUR

Exercise 1, page 107

1. $\frac{1}{6}$
2. $\frac{3}{6}$ or $\frac{1}{2}$
3. $\frac{2}{6}$ or $\frac{1}{3}$
4. $\frac{9}{10}$
5. $\frac{1}{10}$

You Try It, page 108

1. $\frac{1}{4}$ $\frac{\text{(number of fours)}}{\text{(total number of cards)}}$
2. $\frac{3}{4}$ $\frac{\text{(4, 6, and 8 are greater than 3)}}{\text{(total number of cards)}}$
3. $\frac{4}{4}$ or 1 (all numbers are divisible by 2)
4. 0 (none of the numbers is greater than 8)

Exercise 2, page 109

Part One

1. Answers may vary.
2. 1
3. 1
4. Answers may vary.
5. 1
6. 0
7. 1
8. 1
9. 0
10. Answers may vary.

Part Two

Answers may vary. Some examples are given below.

1. By the year 2000, animals will be able to talk as people do.

   A 3-year-old will be elected president in the next election.

   1 + 1 will no longer be equal to 2.
2. Someone in the world will make a new discovery this year.

   A baby will be born in this country this year.

   Some part of the Earth will be in darkness today.

Exercise 3, page 112

1. **Step 1.** $1{,}280 + 960 + 320 = 2{,}560$

   Step 2. $\frac{1{,}280}{2{,}560} = \frac{1}{2}$
2. $\frac{5}{8}$
3. $\frac{1}{8}$
4. less than $\frac{1}{2}$
5. less than $\frac{1}{3}$

Exercise 4, page 115

1. Population: test scores of sixth-graders in state

   Sample: test scores of sixth-graders in six towns
2. Population: low-income people in city

   Sample: people in city low-income housing units
3. Population: defective canisters on early-morning shift

   Sample: defective canisters produced between 12:00 and 1:00 A.M.
4. Population: toy store's merchandise

   Sample: three toys from store
5. Population: all the baked goods in the store

   Sample: chocolate chunk cookies from store

Exercise 5, page 118

1. Population: Americans

   Sample: 200 people living near an army base

   The sample is not representative because people living near an army base may feel differently about defense than others.

2. Population: food at the local markets

   Sample: chickens in five stores

   The sample is not representative of all food in the markets. Where chicken was fresh, produce, for example, may not have been; or vice versa.

3. Population: people of the United States

   Sample: members of American Small Business Owners Association

   The sample is not representative of Americans because it excludes all who do not own a small business.

4. Population: defective circuit boards each month

   Sample: defective circuit boards one Tuesday morning.

   The sample is not representative because the Tuesday morning shift might be the best or worst shift of the week.

You Try It, page 121

Answers should include some of the responses listed below.

1. Mail Advantages: inexpensive; can reach a wide and varied sample

   Disadvantages: low response rate; responses are often not representative of those who received the questionnaire

2. Telephone Advantages: accurate results because of good participation (more representative of population); easy to use

   Disadvantages: leaves out people without telephones; hard to verify information

3. Personal Interview Advantages: reliable; accurate; good participation (more representative of population)

   Disadvantages: expensive; often survey fewer people

Exercise 6, page 122

Answers may vary.

You Try It, page 124

1. 60%
2. 57% to 63%
3. 55%
4. 52% to 58%
5. March: 57%

   April: 58%
6. Yes, when you consider the possible effect of the margin of error on the figures for both months.

Exercise 7, page 125

1. 67% to 75%
2. Yes, it is possible.

   39% + 4% = a possible 43% who said they are happier

   46% − 4% = a possible 42% who said they are not happier

   39% ± 4% = 35% to 43% who said they are happier

   46% ± 4% = 42% to 50% who said they are not happier
3. 525

   35% of 1,500 = 525
4. **a.** 105 (7% of 1,500)

   b. 225 (15% of 1,500)

   c. 150 (10% of 1,500)

CHAPTER FIVE

Exercise 1, pages 136–137

Answers will vary. The following are sample statements.

1. **a.** Although oral contraceptives have a low failure rate in preventing pregnancy, this does not make them the "best" method of birth control. Other factors such as a woman's age, health, lifestyle, and preferences need to be considered, and what is best for one person is not necessarily best for everyone.

   b. Oral contraceptives have a failure rate of 6%, compared to a 16% failure rate for condoms and an 18% failure rate for diaphragms.

2. a. Although North County pays higher police salaries than the other four counties, the salaries might not be *too* high. Maybe North County requires more skills among its police force, or perhaps the other counties' police salaries are too low.

   b. In 1992, full-time North County police officers were paid an average of $30,000 per year, the highest salary among the five counties listed.

3. a. The comparatively large number of people who die in fires in the United States is not necessarily caused by poor fire protection. More fatal fires may occur in the United States because of a more frequent use of chemicals, more flammable building materials, or more multi-story residences.

   b. In the United States in 1991, there were 21 fire deaths per one million people.

Exercise 2, page 141

1. a. Statement 1 refers to Graph **A**.

   b. Statement 2 refers to Graph **B**.

2. **Answers may vary. A sample statement is given below:**

   In 1992, the population in Harris County was approximately 40,000, while in Drake County it was about 70,000.

3. a. Yes

   b. No

4. **Answers may vary. A sample statement is given below:**

   In Graph A, the scale increases by 10,000; in Graph B, the scale increases by 50,000. Because the spacing between 0 and 7 on Graph A is wider than the spacing between 0 and 7 on Graph B, the Graph A bars are longer.

Exercise 3, page 144

1. The statement is not accurate because there is no zero on the scale. The California bar is twice as long as the New Hampshire bar, but the actual birthrates are 20 per 1,000 in California and 15 per 1,000 in New Hampshire.

2.

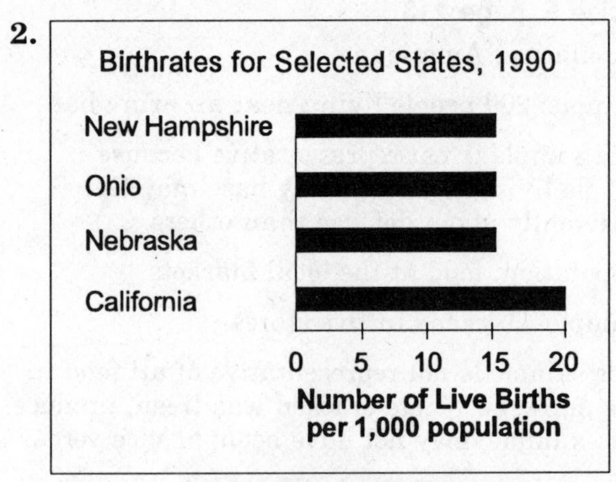

3.

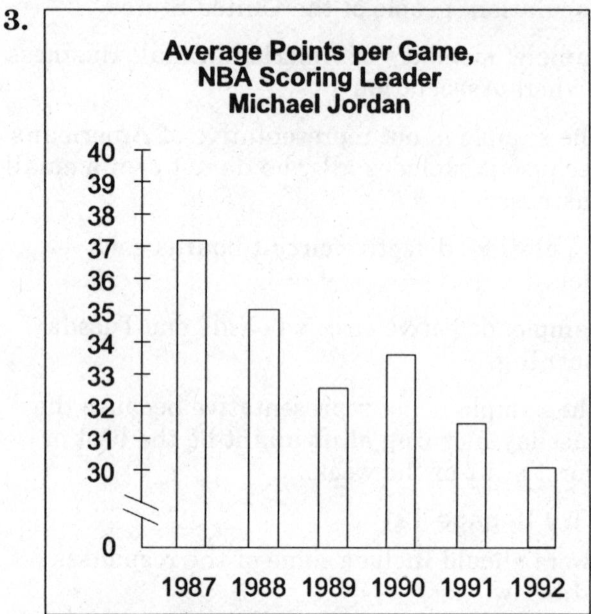

Exercise 4, page 149

1. a. There **is** a correlation between **monthly take-home pay** and **number of televisior sets owned.**

   b. The analysis is not accurate. Although there is a correlation, an increase in one value does not *cause* an increase in the other.

2. a. As the number of school years completed **rises**, the number of hours spent reading per week **rises.** (*Decreases* may also be use in both blanks.)

   b. There **is** a correlation between school years completed and time spent reading.

Exercise 5, page 151

1. **Biased.** The group Citizens for Smith has an interest in more people voting for Smith.

2. **Biased.** A soft-drink company would want more teenagers to prefer its brand.

3. **Biased.** A group called Americans Tough on Criminals would most likely want to find large numbers of people in favor of capital punishment.
4. **Unbiased.** A newspaper would probably have no preference concerning how people feel about gun control.
5. **Biased.** A person trying to get financial support would most likely want to find a large number of people interested in his or her idea.
6. **Unbiased.** A television network generally shows no preference for a particular candidate.

Exercise 6, pages 154–155

1. **Biased.** The question asked is loaded. It emphasizes the advantages of a new factory without mentioning the possible disadvantages.
2. **Unbiased.** The network is most likely looking for facts and has no preference as to how the numbers turn out.
3. **Biased.** This question is loaded, pointing out the disadvantages without mentioning the advantages.
4. **Biased.** The sample surveyed is not representative of Americans in general. People shopping at expensive stores are not likely to be affected by a recession in the same way as other Americans.
5. **Biased.** People leaving an ice cream shop are more likely to prefer ice cream than the general population.

Exercise 7, page 159

1. Answers may vary. A sample statement is given below:

   In 1991, more undocumented female immigrants were stopped by the San Diego Sector of the U.S. Border Patrol than in the previous four years.

2. a

POST-TEST

1. (2) service

   sales: $26 \div 110 = .24$

   service: $10 \div 25 = .4$

2. (2) the percent of people who carpool to other nearby industries
3. Answers will vary. The following is a sample statement:

   The maintenance department has ten times as many employees who carpool to work as management does.

4. (1) The graph shows new home mortgage rates in the United States from 1970 to 1990 in percent per year.
5. (3) 0% to 20%
6. Answers will vary. The following is a sample statement:

   The mortgage rate in 1986 was about the same as in 1990.

7. (1) Selling prices of new homes were higher in 1980 than in 1983.

   The graph shows new home mortgage *rates*, not new home *prices*.

8. (3) American Potato Growers, Inc.
9. (1) Would you prefer a hot, buttered baked potato or a bowl of rice?
10. (3) 3,000 participants at the Potato Lovers Picnic
11. (3) 1 in 5
12. (2) 30–39
13. (2) to show the distribution of AIDS deaths, by age group, from 1982 to 1989
14. Answers will vary. The following is a sample statement:

    More than 1 out of every 100 deaths of children under 12 are AIDS-related.

15. (2) 4.76
16. (3) The chart shows the accident rate per 100 cars for 5 states and the nation in 1989.
17. Answers will vary. The following is a sample statement:

    In 1989, the nation averaged 4.14 accidents per 100 cars, while Massachusetts had more than twice as many.

18. (1) In 1989, Massachusetts had more accidents than any other state listed.

    The chart shows *rates*, not *numbers*.

19. **(2)** 213

Use the graph on the right to find this information.

20. **(3)** As the number of U.S. farms fell from 1940 to 1991, the size of the average farm grew.

The statements in both (1) and (2) are true, but only (3) summarizes the data from both graphs.

21. Answers will vary. The following is a sample statement:

In 1970 in the United States, about 3,000 farms existed, with an average size of 374 acres.

22. **(1)** 1,000,000

There were about 2,000 farms in 1991, with an average size of about 500 acres.

$2{,}000 \times 500 = \mathbf{1{,}000{,}000}$

23. **(1)** yes

As the number of leisure hours rises, the number of medicines taken also rises.

24. **(1)** yes

As age increases, so does the number of medicines taken per year.

25. Answers will vary. The following is a sample summary:

As age increases, so do the number of leisure hours and the number of medicines taken. Of these three factors, age is probably the one that has an effect on the other two. Older people tend to be retired, thus having more leisure time. In addition, older people tend to have more medical problems (high blood pressure, arthritis, etc.) that may require medication.

26. Answers will vary. The following is a sample statement:

The average monthly Calereso commission is projected to be $1,900 in 1993.

27. **(3)** The average monthly commission at Calereso rose steadily between 1990 and 1992.

The 1991 commission is *not* twice the 1990 commission. The graph appears that way because it does not have a zero on the vertical scale.

1994 is not listed on the graph.

28. The 1992 average monthly commission is *not* twice the amount earned in 1990. It appears that way on the graph because there is no zero on the vertical scale.

29. **(2)** 6% to 14%

$10\% - 4\% = 6\%$

$10\% + 4\% = 14\%$

30. **(1)** yes

$46\% + 4\% = 50\% = \frac{1}{2}$

31. **(1)** 200,000

46% is approximately 50%

50% of 400,000 = 200,000

32. **(1)** yes

If Crane actually received 46% − 4%, or 42%, and Davidson received 39% + 4%, or 43%, Davidson would be the winner.

33. **(2)** $\frac{3}{5}$

$60\% = \frac{60}{100} = \frac{60 \div 20}{100 \div 20} = \frac{3}{5}$

34. Answers will vary. The following is a sample statement:

Hispanics in Hunnewell schools make up 5% of the school population, about half the Asian-American population.

35. **(1)** 620

5% of 12,400 = $.05 \times 12{,}400 = 620$ or

10% of 12,400 = 1,240, so 5% of 12,400 = 620